Financial Accounting

with

Microsoft D365 ERP

Satya Kejriwal (C.A.)

Anand Singh Verma (C.A.)

Copyright © 2021 by Satya Kejriwal

Contact:

Email: skejriwal@paretodata365.com
Website: www.satyakejriwal.com

Print ISBN: 979-8474421872
First Edition

INDEX

Game plan to understand this book

Section I: ABC of Accounting

This part is about the fundamentals of finance and accounting. I started studying finance and accounting in school in grade 11 (in 1998) and it took me a few years to understand it in the right context.

I have jotted down the basics of accounting here in a dead-simple way:

- Why your business is different from 'you'?
- Why are most people confused with debit and credit?
- How golden rules of accounting are taught in schools and universities?
- How does the accounting process work behind the scenes?
- *Best practices for designing the chart of accounts*
- **IFRS and International Accounting Standards**

I highly recommend reading this part in detail if you have not read finance & accounting during your school/university time. This part is the building block of this book.

Section II: Microsoft D365 ERP

I recommend this for all D365 consultants, D365 finance users, and new graduates who want to start their D365 careers. It is my promise you will not regret spending time on these 5 chapters.

This part deals with specific accounting with D365 ERP (Finance & Operations). You can also use it as a 'cheat sheet' while configuring the accounting of D365 and there is nothing wrong with that. If you already know a little bit of finance, this section will make you more confident with the core accounting of D365 ERP.

- There is a beautiful link between D365 accounting and Golden Rules of accounting (which you read in Section I)
- How does Microsoft D365 handle accounting for procurement, sales, assets, general ledger, forex, inter-company?
- How the posting works in D365 (which is different from the manual accounting system and differs in each ERP)
- There are some tips and tricks with every chapter which make this book unique.
- IFRS and IAS references at the relevant places.

An example from Chapter 6 below-

IAS 21 — The Effects of Changes in Foreign Exchange Rates

International Accounting Standard (IAS-21) require general ledger account balances in foreign currencies to be revalued using different exchange rate types (current, historical, average, etc.). For example, one accounting convention requires following conversion rates-

- Assets and liabilities→ Current exchange rate,
- Fixed assets→Historical exchange rate, and
- Profit and loss accounts→ Monthly average.

Section III: Financial Statements and Ratio Analysis

It is a bonus section. As you know I am a fan of Warren Buffett and cannot miss financial statements and ratios when we talk about finance and accounting. This part deals with Profit & Loss, Balance Sheet, Cash Flow, etc., and recommends which section and ratio are most important while you want to do a quick health check of a company.

I recommend this for everyone even if you don't want to know about debit-credit or D365 accounting. It will make you a little more intelligent in reading the financials of an organization.

Note- This book is the first in the series of two books. There will be some advanced topics in the next book e.g., revenue recognition, project accounting, lease accounting, landed costing and manufacturing, etc.

Promise from the Authors

This book is an attempt to bring finance and accounting closer to information technology. I had the idea of writing this book while interviewing a candidate for the position of 'finance consultant' in 2015. The consultant was a finance graduate and had 10 years of experience, he was fantastic in ERP configuration but could not answer the basic accounting questions and it was very disappointing for me. That day I decided to write something which can teach accounting even to a 'non-accountant. I strongly believe you will never fail in life if you stick to the basics and build upon that knowledge.

"You have to understand accounting and you have to understand the nuances of accounting. It's the language of business and it's an imperfect language, but unless you are willing to put in the effort to learn accounting— how to read and interpret financial statements— you really shouldn't select stocks yourself." -WARREN BUFFETT

This book covers the basics of accounting along with some advanced topics of D365 and will help all ERP professionals tremendously in their career growth. Just in 10 days, after reading this book-

- You will be more confident while talking to CFOs
- You will never be confused with debit and credit,
- you will know how accounting and posting work in Microsoft D365 Finance.
- You will know some relevant references to IFRS and IAS.
- You will also get an understanding of Profit & Loss, Balance sheet & Cash Flow, ratio analysis which will also help you to analyze the position of a company.

About the Authors

 Satya Kejriwal is a member of the Institute of Chartered Accountants of India (2004 batch). He started his career in finance after graduation but his passion for IT pulled him into ERP consulting soon. In the last 15 years, Satya has worked with IL&FS Technologies, Hitachi Solutions, Breville, Microsoft. He has been part of several large D365 ERP (F&O) projects in India, South-East Asia, America, and the ANZ region before starting his ERP consulting in 2019. Satya is passionate about transformation and a strong believer of Pareto Analysis (80:20 Rule) in life. He is also an avid value investor and a big fan of Warren Buffett.

Satya is settled in Sydney and his personal life, he is married, and his wife works with a reputed organization in Sydney as a CRM consultant. They have a 10-year-old son. Satya is crazy about sports, yoga, travelling, and singing.

 Anand Singh Verma is a member of the Institute of Chartered Accountants of India. (2006 batch). He has been associated in different capacities with *Ernst & Young (E&Y) & other companies* before he started his practice as 'ASV & Co, Chartered Accountants'. He is IFRS qualified from ICAI and has worked extensively in auditing, corporate structuring, entity/share valuation of large domestic and multinational companies, outsourcing of multinational companies having their subsidiaries/project office/branches in India.

In personal life, Anand is married and has two beautiful kids (Swasti and Swastik). Anand is a nature loving person and has a start-up in organic food as well. His wife is a popular yoga & meditation instructor in India.

Foreword

Microsoft Dynamics 365 for Finance and Operations may be a new name, it has been around for more than two decades. Microsoft Dynamics 365 for F&O is a modern/next-generation ERP system built on cloud-first fundamentals and with two decades of carefully curated proven business functionalities and in the past known as Microsoft Dynamics AX and Axapta.

Personally, I have been associated with this product for over two decades and witnessed it become the most comprehensive enterprise software in the market, which brings together intuition, intelligence, and mobility to users and facilitates true digital transformation for businesses of all sizes.

Organizations across the world right now will be thinking about more significant shifts in the way they deliver their products and services to their end customers. It doesn't really matter in which industry—As business complexity rises, organizations increasingly turn to their senior finance leaders for the strategic direction and to drive business transformation. Chief financial officers (CFOs) play significant roles in companies' transformation efforts, they must look for opportunities to leverage technology, to enhance financial decision-making within their organizations. Moreover, by projecting future outcomes, organizations can make more informed decisions today.

We've seen incredible growth of Dynamics 365 just in the past year. This momentum is driving massive investment in people "For Dynamics professionals & "For those looking to get into Dynamics, I cannot think of a better time to make a difference by helping organizations digitally transform.

Having said that the book *'Financial Accounting with Microsoft D365 ERP'* is a *'back-to-basics'* book and the first in the series of two books. *The authors have covered not only D365 accounting in simple words, but also there is special coverage of International Financial Reporting Standards (IFRS) and International Accounting Standards (IAS).*

Accounting standards worldwide are shifting to or converging with IFRS. For enterprises affected by this shift, it will likely require changes and modifications to the ERP systems used to collect and report financial data. The movement to IFRS can impact key processes and systems in treasury, payroll, general ledger, financial instruments, and asset management, to name a few. If companies have started or are about to start an ERP or finance transformation project, now could also be the time to factor in IFRS considerations. D365 ERP has been designed to incorporate IFRS. Thus, a finance transformation project conducted in conjunction with an IFRS assessment that effectively maps IFRS-related changes into the ERP can produce efficiencies for both the initiatives, which could be very significant.

I know Satya Kejriwal, the author of the book for the last few years now. He has been one of the leaders in D365 ERP and is known for his acumen in ERP implementation. Who could write such a book better than him as he is a chartered accountant and has spent more than 15 years with this ERP?

I will highly recommend this book for everyone who is working with D365 F&O or wants to get into D365 Finance and experience the carefully curated proven business functionalities of this next-generation ERP.

Cheers.

Shashidhar Dayaneni
Managing Architect (ASIA)
Microsoft Industry Solutions Delivery

Section-I
ABC of Accounting

1
Introduction

I would like to start the introduction with a quote from my favourite investor on the earth

*"You have to understand accounting and you have to understand the nuances of accounting. It's the language of business and it's an imperfect language, but unless you are willing to put in the effort to learn accounting— how to read and interpret financial statements— you really shouldn't select stocks yourself." — **WARREN BUFFETT***

As Warren said above, accounting is the language of businesses. Stock markets all over the world use this language to communicate financial transactions and results.

The origin of accounting is as old as money. In the early days, the number of transactions was small, so every concerned person could keep a record of transactions during a specific period. Gradually, the field of accounting has undergone remarkable changes in compliance with the changes happening in the business scenario of the world.

An accountant records transactions according to certain accounting principles and standards depending upon the size, nature, volume, and other constraints of a particular organization.

I apologize ☺ for starting this book with this chapter which you might find boring and overwhelming if you are new to the accounting world.

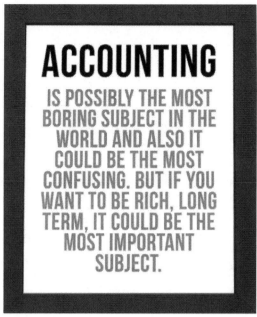

If the above statement motivates you, continue reading this book, stay focused for the next 10 minutes, give it a try and proceed to the next chapter with whatever sticks to your mind. I will be using references to the 'Accounting processes' from this chapter throughout the book.

Definition of Accounting

The American *Institute of Certified Public Accountant* has defined Financial Accounting as:

*"The art of **recording, classifying and summarizing** in a significant manner and in terms of money, transactions and events which in part at least of **a financial character** and interpreting the **results** thereof."*

Objectives and Scope of Accounting

Let us go through the main objectives of accounting:

- **To keep systematic records**

- **To ascertain profitability**

- **To ascertain the financial position of the business**

- **To assist in decision-making**

- **To fulfill compliance of Law**

All the above points are self-explanatory, so we will now quickly move on from here.

Accounting Process

Accounting involves certain steps which are repeated again and again during the accounting cycle. The length of an accounting cycle can be monthly, quarterly, half-yearly, or annually. It may vary from organization to organization, but the

> 🔅 Listed companies all over the world follow quarterly accounting cycle to report the results in stock market but generally everyone closes their books every month.

process remains the same.

The following chart shows the basic steps in an accounting cycle:

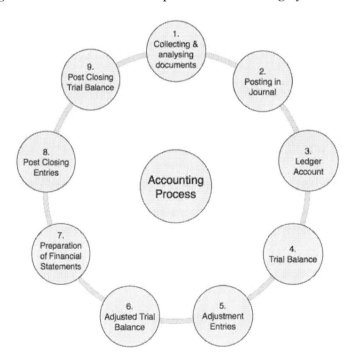

The following table lists down the steps followed in an accounting process:

1. Collecting and Analyzing Accounting Documents

2. Posting in Journal

3. Posting in Ledger Accounts

4. Preparation of Trial Balance

5. Posting of Adjustment Entries

6. Adjusted Trial Balance

7. Preparation of Financial Statements

8. Post-Closing Entries

9. Post-Closing Trial Balance

We will talk about the process in detail in **chapter-3** and compare how this process works in ERPs.

What is the difference between finance & accounting?

95% of the people don't know the difference between finance and accounting. The terms 'Finance' & 'Accounting' are always used together, and it is hard to separate these two roles in an organization. There is a very thin line between these two.

- Most of the companies have a finance department and the accounting section falls under the finance umbrella.

- Some smaller companies don't make much difference in finance and accounting, they are considered synonyms.

Definition of Finance: -
*"The art of **managing money and its investment** with the assurance that the funds are allocated among the various assets classes **to reap the maximum potential return.**"*

Having said that there is a fundamental difference between these two terms or between two roles.

Accounting
What happened in the past?

Finance
What will happen in the future?

	Accounting	Finance
Definition	"The art of **recording, classifying, and summarizing** in a significant manner and in terms of money, transactions and events which in part at least of **a financial character** and interpreting the **results** thereof."	"The art of **managing money and its investment** with the assurance that the funds are allocated among the various assets classes **to reap the maximum potential return.**"
What is their approach?	It is based on the past data and record/analyzes **what happened in the past**	It is futuristic and makes the **decision for future**
Responsibilities	Accounting & Tax Compliance	Forming strategies for creating value for the investor
What subjects they should know? (Examples)	Generally, **CPA, Chartered Accountants, Cost Accountants** are hired for this role since they are the experts in these subject areas: • Auditing • Taxation	There are specialized **MBA** courses for finance, but **CAs & CPAs** are also known to master finance. They should know these subjects: • Corporate Finance • Derivatives

	• GAAP/IFRS • Management Accounting • Cost Accounting	• Capital Markets • Portfolio Management • Financial Modelling
What is their skill set?	• Detail & Procedure oriented • Well informed on applicable statutes & accounting rules • Quantitative and computation skills	• Research-oriented & analytical inclination • Well informed of current market situations • Result & profit-driven
What do they produce? (examples)	Accounting is responsible for • Book-keeping • Balance Sheet • Profit & Los • Cash Flow Statements • Budgets	Finance includes • Capital Budget • Working Capital Budget • Investment Policy • Forecasting schedules • Leverage Analysis
Whom do they help in business?	The accounting team actively supports the finance team by giving accurate and relevant data for the company	The finance team sets the pace for business by guiding them on the targets to be chased and removing any money related hurdles

As you saw in the comparison above, finance is more future-oriented, and it is an art and a personal skill that you learn with experience. There are tools and techniques and data in the market but ultimately how you apply in the real life, comes by experience. This is the reason people from accounting find it easy to transition to finance roles as they have a strong base of accounting and taxation, and they spend enough time in the business and learn finance on the job.

You must be curious now. Are we dealing with finance or accounting or both in this book? I will be honest in saying that 80% of this book is accounting and 20% is finance.

1A

Basic Principles of Accounting

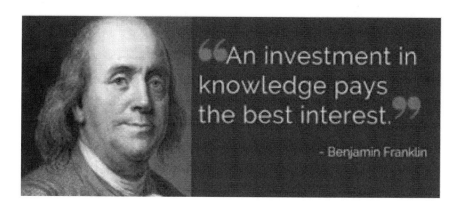

Accounting Concepts
- Business Entity
- Money Measurement
- Going Concern Concept
- Dual Aspect Concept
- Accounting Period Concept
- Matching Concept
- Accrual Concept
- Object Evidence Concept

Accounting Conventions
- Consistency
- Disclosure
- Materiality
- Prudence

Accounting Concepts & Conventions

> When I started reading accounting in school, I realized these 'Accounting concepts' and 'Accounting Conventions' are the foundation of accounting. So, read these now and I strongly recommend you coming back to these concepts when you are on chapter 3.

(A) Accounting Concepts

There are some basic concepts of accounting that are followed globally and <u>will never change in the future-</u>

- ➤ Business Entity Concept
- ➤ Money Measurement Concept
- ➤ Going Concern Concept
- ➤ Dual Aspects Concept
- ➤ Accounting Period Concept
- ➤ Matching Concept
- ➤ Accrual Concept
- ➤ Objective Evidence Concept

The first two accounting concepts, namely, Business Entity Concept and Money Measurement Concept are the fundamental concepts of accounting. Let us go through each one of them briefly:

Business Entity Concept

According to this concept, the business and the owner of the business are two different entities. In other words, I and my business are separate.

For example, Mr. Bill starts a new business in the name of Microsoft Corp and introduced a capital of $2,00,000 in cash. It means the cash balance of Microsoft Corp will increase by a sum of $2,00,000/-. At the same time, the liability of Microsoft Corp in the form of capital will also increase. It means Microsoft Corp. is liable to pay $2,00,000 to Mr. Bill.

Money Measurement Concept

According to this concept, "we can book only those transactions in our accounting record which can be measured in monetary terms."

Example

Determine and book the/ value of stock of the following items:

Shirts	$ 5,000/-
Pants	$ 7,500/-
Coats	500 pieces
Jackets	1000 pieces
Stock Value	???

Here, if we want to book the value of stock in our accounting record, we need the value of coats and jackets in terms of money. Now if we conclude that the values of coats and jackets are $ 2,000 and $ 15,000 respectively, then we can easily book the value of the stock as $ 29,500 (as a result of 5000+7500+2000+15000) in our books. We need to keep quantitative records separately.

Going Concern Concept

Our accounting is based on the assumption that a business unit is a going concern which means that the business will run for a long time. We record all the financial transactions of a business in keeping this point of view in our mind that a business unit is a going concern, not a gone concern. Otherwise, the banker will not provide loans, the supplier will not supply goods or services, the employees will not work properly, and the method of recording the transaction will change altogether.

For example, a business unit makes investments in the form of fixed assets, and we book only depreciation of the assets in our profit & loss account, not the difference of acquisition cost of assets less net realizable value of the assets. The reason is simple; we assume that we will use these assets and earn profit in the future while using them. Similarly, we treat deferred revenue expenditure and prepaid expenditure. The concept of going concern does not work in the following cases:

➢ If a unit is declared sick (unused or unusable unit).

➢ When a company is going to liquidate, and a liquidator is appointed for the same.

➢ When a business unit is passing through a severe financial crisis and going to wind up.

Dual Aspect Concept

There must be a double entry to complete any financial transaction, which means debit should be always equal to the credit. Hence, every financial transaction has its dual aspect:

- ➢ we get some benefit, and
- ➢ we pay some benefit.

For example, if we buy some stock, then it will have two effects:

- ➢ the value of the stock will increase (get the benefit for the same amount), and

- ➢ it will decrease the cash balance if you pay in cash OR increase our liability in the form of creditors.

	Stock will increase by $ 25,000 (Increase in debit balance)
Purchase of Stock for $25000	Cash will decrease by $ 25,000 (Decrease in debit balance) or Creditor will increase by $ 25,000 (Increase in credit balance)

We have a big discussion in Chapter-3 on the very topic, so don't get intimidated if you don't understand it now.

Accounting Period Concept

The life of a business unit is indefinite as per the going concern concept. To determine the profit or loss of a firm, and to ascertain its financial position, profit & loss accounts and balance sheets are prepared at regular intervals of time, usually at the end of each year. This one-year cycle is known as the accounting period. The purpose of having an accounting period is to take corrective measures keeping in view the past performances, to nullify the effect of seasonal changes, to pay taxes, etc.

Based on this concept, revenue expenditure and capital expenditure are segregated. Revenue's expenditures are debited to the profit & loss account to ascertain correct profit or loss during a particular accounting period. Capital expenditure comes in the category of those expenses, the benefit of which will be utilized in the next coming accounting periods as well.

The accounting period helps us ascertain the correct position of the firm at regular intervals of time, i.e., at the end of each accounting period.

Matching Concept

The matching concept is based on the accounting period concept. The expenditures of a company for a particular accounting period are to be matched with the revenue of the same accounting period to ascertain accurate profit or loss of the firm for the same period. This practice of matching is widely accepted all over the world. Let us take an example to understand the Matching Concept clearly.

The following data is received from M/s Globe Enterprises during the period 01-01-2020 to 31-12-2020:

Particulars	Amount
1. Sale of 1,000 Electric Bulbs @ $ 10 per bulb on cash basis.	10,000.00
2. Sale of 200 Electric Bulb @ $. 10 per bulb on credit to M/s Atul Traders.	2,000.00

3. Sale of 450 Tube light @ $.100 per piece on Cash basis.	45,000.00
4. Purchases made from XZY Ltd.	40,000.00
5. Cash paid to XYZ Ltd.	38,000.00
6. Freight Charges paid on purchases	5,000.00
7. Electricity Expenses of shop paid	1,500.00
8. Bill for Dec-2020 for electricity still outstanding to be paid next year.	1,000.00

Based on the above data, the profit or loss of the firm is calculated as follows:

Particulars	Amount	Total
Sale		
Bulb	12,000.00	
Tube	45,000.00	57,000.00
Less: -		
Purchases	40,000.00	
Freight Charges	5,000.00	
Electricity Expenses	1,500.00	
Outstanding Expenses	1,000.00	47,500.00
Net Profit		9,500.00

In the above example, to match expenditures and revenues during the same accounting period, we added the credit purchase as well as the outstanding expenses of this accounting year to ascertain the correct profit for the accounting period 01-01-2020 to 31-12-2020.

It means the collection of cash and payment in cash is ignored while calculating the profit or loss of the year. It is also as accrual accounting.

Accrual Concept

As stated above in the matching concept, the revenue generated in the accounting period is considered and the expenditure related to the accounting period is also considered. Based on the accrual concept of accounting, if we sell some items or rendered some service, then that becomes our point of revenue generation irrespective of whether we received cash or not. The same concept is applicable in the case of expenses. All the expenses paid in cash or payable are considered and the advance payment of expenses, if any, is deducted.

Objective Evidence Concept

Every financial entry should be supported by some evidence. The purchase should be supported by purchase bills, sale with sale bills, cash payment of expenditure with cash memos, and payment to creditors with cash receipts and bank statements. Similarly, the stock should be checked by physical verification and the value of it should be verified with purchase bills. In the absence of these, the accounting result will not be trustworthy, chances of manipulation in accounting records will be high, and no one will be able to rely on such financial statements.

(B) <u>Accounting Conventions</u>

We will discuss the following accounting conventions in this section:

- ✓ Convention of Consistency
- ✓ Convention of Disclosure
- ✓ Convention of Materiality
- ✓ Conservation of Prudence

Convention of Consistency

To compare the results of different years, it is necessary that accounting rules, principles, conventions and accounting concepts for similar transactions are followed consistently and continuously. The reliability of financial statements may be lost if frequent changes are observed in accounting treatment. For example, if a firm chooses *cost or market price whichever is lower* method for stock valuation and *written down value method* for depreciation to fixed assets, it should be followed consistently and continuously.

Consistency also states that if a change becomes necessary, the change and its effects on profit or loss and on the financial position of the company should be clearly mentioned.

Convention of Disclosure

IFRS, IAS and local laws of each country prescribe a format in which financial statements must be prepared. The purpose of these provisions is to disclose all essential information so that the view of financial statements should be true and fair. However, the term 'disclosure' does not mean all information. It means disclosure of information that is significant to the users of these financial statements, such as investors, owners, and creditors.

Convention of Materiality

If the disclosure or non-disclosure of information might influence the decision of the users of financial statements, then that information should be disclosed.

Conservation or Prudence

It is a policy of playing safe. For future events, profits are not anticipated, but provisions for losses are provided as a policy of conservatism. Under this policy, provisions are made for doubtful debts as well as a contingent liability; but we do not consider any anticipatory gain.

For example, If A purchases 1000 items @ $ 80 per item and sells 900 items out of them @ $ 100 per item when the market value of a stock is (i) $ 90 and in condition (ii) $ 70 per item, then the profit from the above transactions can be calculated as follows:

Particulars	Condition (i)	Condition (ii)

Sale Value (A)	90,000.00	90,000.00
(900 x 100)		
Less:- Cost of Goods Sold		
Purchases	80,000.00	80,000.00
Less Closing Stock	8,000.00	7,000.00
Cost of Goods Sold (B)	72,000.00	73,000.00
Profit (A-B)	18,000.00	17,000.00

In the above example, the method for valuation of stock is 'Cost or market price whichever is lower'.

The prudence however does not permit the creation of hidden reserve by understating the profits or by overstating the losses.

Importance of Accounting for Entrepreneurs, Businessmen, Managers

Accounting is the most important part of any successful business. It records all profits, losses, credits, assets, and debts. It tells you the state of the business in numbers, not words. It provides the most vital information you need to understand how your business grows, makes money, where the profit of a business goes, and what your cash flow is. In short, if you do not understand the basic principles of accounting, you cannot run a business, nor can you even hope to help a business grow and profit.

Today in the world, we face a new economic era where, day by day, labor-driven jobs disappear, being replaced by an ever-increasing number of better-paying, general management jobs. If you work for a company and are currently managing any aspect of the company, or if you aspire to move up to management, then you need to understand what accounting is. Moreover, if you are an entrepreneur, or you ever plan to start your own business, you need to understand, at the very least, the basic principles of accounting.

Many times, people hired into high-ranking positions in the corporate world have no concept of basic accounting. In fact, at the mere mention of the word "accounting", they become withdrawn and quiet. Yet these same managers are somehow going to increase sales and overall profit of a company.

Learning accounting is like any new skill. There is a learning curve, and the skill needs to be practiced, or used in this case, for it to be effective. If you have access to your company's financial statements, please take the time to apply our examples to your company's financials.

Those of us who are born as right-brain thinkers tend to be better at creative, imaginative, passionate activities. Then there are those of us who are left-brain thinkers, naturally adept at working with numbers, applying logical reasoning, and analytically solving problems.

Regardless of which category you fall into, the truth is that anyone can learn the basic principles of accounting and develop a knack for managing the financial aspects of a business. The upside of learning basic accounting principles is that, regardless of whether it is a large Fortune 500 company or a small entrepreneurial start-up, the same fundamental rules apply when working with the bottom line.

2
Debits and Credits: Change Your Paradigm

Overview of Debit and Credit

Traditionally, the posting of debit and credit transactions has been called Accounting or Bookkeeping. But the broader term, Accounting, is now commonly used to include the recording of financial transactions for a company. For help with the most common transactions, a small business might enter an accounting software or ERP system.

Most accounting and bookkeeping software, such as QuickBooks or Xero or Tally, are marketed as easy to use. But if you don't know some bookkeeping basics, you will make mistakes. If you never "kept books" manually, reading a phrase such as "debits always go on the left and credits always go on the right" brings no joy. I will explain debits and credits in a new way - using basic math concepts! But before we jump into that, understand the following two things: -

A. Five types of Accounts

B. Classification of Accounts

(A) The Five types of Accounts in an Accounting System:

To fully understand how to record bookkeeping transactions, we must understand that all our accounts fit into one of 5 categories. The account categories are:

A. **Assets:** what the company owns of value (Cash, Accounts Receivable, furniture, vehicles)
B. **Liabilities:** what the company owes to others (loans, Accounts Payable)
C. **Equity:** the company's net worth. Equity equals Assets minus Liabilities.
D. **Revenue:** money the company is earning
E. **Expenses:** money the company is spending

(B) Classification of Accounts

It is necessary to know the classification of accounts and their treatment in double entry system of accounts. Broadly, the accounts are classified into three categories:

1. Personal accounts
2. Real accounts
   - Tangible accounts
   - Intangible accounts
3. Nominal accounts

Let us go through them each of them one by one.

1. **Personal Accounts**

   Personal accounts are the accounts that represent an individual person or an organization in the business. These can be classified in the following categories:

   a) **Natural Personal Account: -** An account related to any individual like David, George, Satya or Sandeep is called as a *Natural Personal Account.*

   b) **Artificial Personal Account: -** An account related to any artificial person like ABC Ltd, M/s Reliance Industries, Amazon Co. Limited etc. are called as an *Artificial Personal Account.*

c) **Representative Personal Account:** - Representative personal account represents a group of account. If there are several accounts of similar nature, it is better to group them like salary payable account, rent payable account, insurance prepaid account, interest receivable account, capital account.

> -☼- All the customers and vendors in a business can be called personal accounts. Accounts receivable and Accounts Payable modules of an accounting system are all about handling the 'Personal Accounts'. We will understand in the next chapter the importance of personal accounts and their treatment in the accounting world.

2. Real Accounts

Every Business has some asset, and every asset has an account. Thus, an asset account is called a real account. There are two types of assets:

Tangible assets are touchable assets such as plants, machinery, furniture, stock, cash, etc.
Intangible assets are non-touchable assets such as goodwill, patent, copyrights, etc.
Accounting treatment for both types of assets is the same.

> -☼- The Fixed Asset Module of an accounting system deals with the real accounts.

3. Nominal Accounts

Since this account does not represent any tangible asset, it is called a nominal or fictitious account. All kinds of an expense account, loss account, gain account, or income accounts come under the category of nominal account. For example, rent account, salary account, electricity expenses account, interest income account, etc.

> -☼- All the accounts in a 'Profit & Loss' statement are called nominal accounts.

The Best strategy to understand Debit-Credit for beginners

Golden Rules of Accounting

The following rules of debit and credit are called the golden rules of accounts:

Rule#	Rules	Applicable on	Net Effect
1	Debit what comes in Credit what Goes Out **OR** Debit the increase in the assets/Credit the decrease in the assets. Credit the increase in the liabilities/Debit the decrease in the liabilities	Personal Accounts & Real Accounts	Debit = credit
2	Debit all the Expenses/Losses. Credit All the Incomes/Profit	Nominal Accounts	Debit = credit

	ASSET	EXPENSE	LIABILITY	INCOME	EQUITY
DEBIT	↑	↑	↓	↓	↓
CREDIT	↓	↓	↑	↑	↑

Generally, these rules are used in combination. Let's understand this with help of the following story of Mr. George who is based in Hongkong and starts a furniture business. He floats a company named ABC Furniture Ltd. and invests some money in the business along with some of the other investors.

☀ You should take the print of these golden rules and read it daily unless you are an expert in accounting.

Transaction#1 Introduction of Share Capital

When the investment money is deposited into the bank by the investors-

Debit/ Credit	Type of Account	Account details	Class of Account	Rule#	Description
Dr.	Asset	Bank A/c	Real	#1	Debit what comes in / Debit the increase in the assets
Cr	Capital	Share Capital A/	Personal	#1	Credit the increase in the liability

✓ *The bank account is debited* with Rule#1 since the bank is a real account and it will be debited if it comes in or increases in value.

✓ *Why is the 'Share Capital A/c' credited?* Think from the business perspective (refer to 'Business Entity Concept' from chapter-1A), the investors are independent of the business, and they are lending money to the business and hence they are the creditors or liability. Going by rule#1, we should credit the increase in the liability.

Transaction#2 Purchase of Fixed Assets

ABC furniture needs some assets (e.g., computers, machines) in the back office and factory to manufacture the furniture or keeping the books. So, the purchasing department buys some assets from the vendors. (We are buying machinery for factory and furniture for office in this example)

Debit/ Credit	Type of Account	Account details	Class of Account	Rule#	Description
Dr.	Asset	Machinery & Equipment A/c	Real	#1	Debit the increase in the assets
Cr.	Liability	Dell computers (AP) A/c	Personal	#1	Credit the increase in the liability
Dr.	Asset	Furniture & Fixture A/c	Real	#1	Debit the increase in the assets

Cr.	Liability	Ikea Furniture	Personal	#1	Credit the increase in the liability

✓ Machinery or furniture accounts are real accounts so rule#1 will be applicable.

✓ Dell computers or Ikea Furniture are personal accounts, so rule#1 will be applicable.

Transaction#3 Payment to the non-trade suppliers

The invoice of the Ikea furniture must be paid immediately, the amount is transferred from the bank to the supplier.

Debit/ Credit	Type of Account	Account details	Class of Account	Rule#	Description
Dr.	Liability	Ikea Furniture	Personal	#1	Debit the decrease in the liability
Cr.	Asset	Bank A/c	Real	#1	Credit the decrease in the asset

Transaction#4 Purchase of Stock

Now, there is some furniture which is bought for trading purposes since ABC Furniture Ltd. is a furniture company.

Debit/ Credit	Type of Account	Account details	Class of Account	Rule#	Description
Dr.	Asset	Inventory A/c	Real	#1	Debit the increase in the assets
Cr.	Liability	Vendor (AP) A/c	Personal	#1	Credit the increase in the liability

✓ The accounting treatment is almost the same in this case also except the nature of the real account is 'Inventory account' instead of 'Furniture & Fixture Account.'

Transaction#5 Freight on Stock paid

Payment made to the transporter who brought the furniture to the warehouse

Debit/ Credit	Type of Account	Account details	Class of Account	Rule#	Description
Dr.	Expense	Freight A/c	Nominal	#2	Debit all the expenses/losses
Cr.	Asset	Bank A/c	Real	#1	Credit the decrease in the assets

✓ Freight is a nominal account, so going by rule#2, it is debited here and

---

🔅 Instead of debiting the 'freight A/c', we could also debit the 'Stock A/c' since it is adding the value of the stock. But don't worry about this treatment yet. Just focus on the basic accounting for now.

✓ Bank which is a real account, and is decreasing, so it is credited here.

Transaction#6 Sale of stock to a customer

A customer orders 10 chairs. Following is the accounting entry for a sale transaction-

Debit/ Credit	Type of Account	Account details	Class of Account	Rule#	Description
Dr.	Asset	Customer (AR) A/c	Personal	#1	Debit the increase in the assets
Cr.	Revenue	Sales A/c	Nominal	#2	Credit all income/profit
Dr.	Expense	Cost of	Nominal	#2	Debit all the

		Goods Sold			expenses/losses
Cr.	Asset	Inventory A/c	Real	#1	Credit the decrease in the assets

This transaction is a bit complex for beginners but let's try to decode this also in two parts. Part-I

- ✓ Customer account will be debited because he is an asset since we need to recover money from him after we sold goods to him on credit.

- ✓ sales account (revenue) account will be credited (rule#2) since it is the income of the business.

Part-II

- ✓ Inventory account will be credited with the cost amount (rule#1) since it is real account, and it is decreasing

- ✓ Debit side of this transaction will be 'Cost of Goods Sold (COGS)' which means that we are converting that inventory into our cost now and all expenses are debited.

Transaction#7 Collection from the customer

The customer makes the payment for the purchase he made from ABC

Debit/ Credit	Type of Account	Account details	Class of Account	Rule#	Description
Dr.	Asset	Bank A/c	Real	#1	Debit the increase in the assets
Cr.	Asset	Customer (AR) A/c	Personal	#1	Credit the decrease in the assets

- ✓ Since the bank balance will increase, it will be debited and

- ✓ the customer account (an asset) is decreasing, and it will be credited as per rule#1.

It is very clear from the above examples how the rules of debit and credit work. It is also clear that every entry has its dual aspect. In any case, the debit will always be equal to credit in double-entry accounting system.

> **⚠ Caution:** Golden rules are the building block of accounting or book-keeping. The road ahead is a bit slippery, so don't move further to chapter-3 if you are not crystal clear with these rules. Send me an email if you get stuck on this, we can talk, and I will try my best to clarify this for you.

Why most people are confused about debit and credit.

One reason many folks are confused about debits and credits is that they believe that credits mean that they are "receiving money." You return an item to the store and you receive a store credit, right? Or the store may "credit" your charge card - giving money back to you.

These are all true ... but here is the big problem: *These stores and banks are using the term "credit" from their perspective!* In other words, when the store or bank gives you a credit, it is *their Cash that they are crediting!* The bank is subtracting money from *their* Cash and giving it to you. As a business owner, you must think of debits and credits from your company's perspective. **When you credit Cash, you subtract from it. Likewise, when you debit Cash, you add to it.**

> 💡 And another little fact you should know accountants and bookkeepers often use DR (debit record) to indicate a debit, and CR (credit record) to indicate a credit.

Debit and Credit-Conclusion

If you fully understand the above, you will find it much easier to determine which accounts need to be debited and credited in your transactions. Modern accounting software helps us when it comes to Cash. When you enter a deposit, most software such as 'Microsoft D365 Finance' automatically debits 'Cash' so you just need to choose which account should receive the credit. And when writing a check, the software *automatically credits Cash*, so you just need to select the account to receive the debit (perhaps an Expense account or a supplier account).

Test your knowledge.

Test your knowledge with the following questions: -

1. what will be debited when you purchase inventory from a party?

    A. Stock

    B. Vendor

    C. Customer

    D. Expense

2. What will be debited when you sell to a customer?

    A. Sales

    B. Vendor

    C. Customer

    D. Income

3. What account will be debited when an amount is received from a customer?

    A. Income

    B. Customer

    C. Ledger

    D. Bank

4. What account will be credited when you purchase Printing & Stationery from a supplier?

    A. Expense

    B. Vendor

C. Customer

D. Bank

5. What account will be debited when you enter bad debts for a debtor

A. Provision for bad debts

B. Bad debts

C. Customer

D. Vendor

6. What account will be credited when you receive Interest from a customer

A. Interest received

B. Customer

C. Cash

D. Bank

7. What account will be credited when the insurance amount is paid for the whole year to an insurance company

A. Prepaid insurance

B. Vendor

C. Expense

D. Insurance

8. What account will be debited when carriage outward is paid in cash

A. Carriage outward

B. Vendor

C. Expense

D. Cash

9. What account will be debited when commission is paid to an agent

   A. Commission A/C

   B. Commission payable

   C. Agent

   D. Bank/Cash.

10. What account will be credited when salary is paid to an employee

A. Salary

B. Salary payable

C. Cash/Bank

D. Employee

Visit our website www.satyakejriwal.com for the answers. You will find more questions to practice there.

3

Accounting Process

Before we proceed to the accounting process or the nitty-gritty of the accounting, you should be very clear with our objectives of the accounting. What is the whole purpose of maintaining accounting books, chart of accounts, debit/credit, etc?

The primary purpose of accounting is to measure the performance of a business in terms of numbers. Have you ever heard about the term 'DMAIC'? No? This abbreviation stands for 'Define, measure, analyze, improve, control' and it is widely used in the six-sigma management philosophy. This concept was originated for the quality control process, but we can also apply it in any business, sports, daily life. Let's understand a simple example-

Three years back, when I started training for a marathon, I could run at the speed of 7 km/hour, so I did DMAIC for my project which I named 'Go-Marathon'.

#	Steps	Go-Marathon
1	Define	Define the target speed you want to achieve e.g. 12km/hour
2	Measure	Measure your current performance e.g. 7km/hour
3	Analyze	Deviation of 5km/hour
4	Improve	Start training, improve diet & lifestyle, changing shoes
5	Control	Measure again after a month and repeat steps 2 to 4

It is nothing unusual, everyone does the same thing but generally, we don't do it in a set sequence or in a disciplined way.

Now, if we apply the same concept to the furniture business of Mr. George, the steps will exactly be the same, only the improvement actions will be unique to your business.

#	Steps	Business	Business term
1	Define	Define the profit you want to achieve in the business e.g. $10Million	Budget or Target, Define KPI
2	Measure	Measure your current performance e.g. $7Million	Profit & Loss, Balance Sheet, Cashflow, Ratio analysis for measuring KPI
3	Analyze	Deviation of $3Million	Budget vs. Actual
4	Improve	Start training your sales team, cost-cutting, improve products,	Quality improvement, sales

		etc.	training, etc.
5	Control	Measure again after a month and repeat steps 2 to 4	Budget control, quality control, etc.

Now coming back to the purpose of accounting, any business starts from defining its financial goals and periodically measuring its financial state. Generally, the accounting software helps in step#2 (recording the transactions and measuring the results) but the latest ERP systems in the market take care of steps #1, 2, and 3 i.e., you can define budget, record actual transactions, and analyze budget vs actual.

> There is a very famous saying in six-sigma technology "If you can define it, you can measure it. if you can measure it, you can control it"

So, you must have got a sense of the importance an accounting system has for a business. Now, let's check on the basic components and terminology you need to understand in accounting-

- Balance sheet

- Profit & Loss

- Cashflow

- Chart of accounts

- Ledger

- Sub-ledger

- Journals

These are not the only terms or components in accounting, but these are building blocks. If you can understand these, you are on the path to mastery in accounting. Trust me!

Nobody will teach you these basic things which are generally taught in high school or college. And ironically, this is the minimum level of knowledge is expected from you when you meet with clients as a consultant.

Accounting Cycle

The accounting cycle refers to the steps required in completing an accounting process. The length of an accounting cycle can be monthly, quarterly, half-yearly, or annually. It may vary from organization to organization, but the process remains the same. *You must have heard that the accounting team is busy doing their month closing and sometimes year closing, so basically, they are following monthly and the yearly accounting cycle.* At the end of each accounting cycle, they are supposed to generate the financial statements (e.g., trial balance, balance sheet, profit & loss, cash flow, etc.) The following chart shows the basic steps in an accounting cycle:

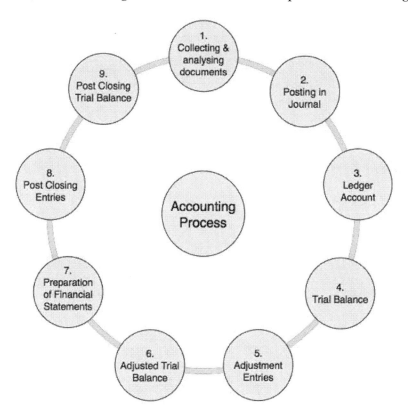

Steps	Manual Accounting	Accounting Cycle in ERP

1. Collecting and Analyzing Accounting Documents	It is a very important step in which you examine the source documents and analyze them. For example, cash, bank, sales, and purchase-related documents. This is a continuous process Throughout the accounting period.	Collecting and analyzing the accounting documents is an offline process and must be done manually.
2. Posting in Journal	Based on the above documents, you pass journal entries using a double entry system in which debit and a credit balance remain equal. This process is repeated throughout the accounting period.	Several journals are used for posting based on transaction type e.g *General Journal, Vendor Invoice Journal, Purchase Order, Sales order, Free text invoice, Payment Journal, etc.*
3. Posting in Ledger Accounts	The debit and credit balance of all the above accounts affected through journal entries are posted in ledger accounts. A ledger is simply a collection of all accounts. Usually, this is also a continuous process for the whole accounting period.	Posting in ledger accounts is an automatic process in 'Dynamics 365' as soon as the journals are posted in the system.
4. Preparation of Trial Balance	As the name suggests, a trial balance is a summary of all the balances of ledger accounts irrespective of whether they carry a debit balance or credit balance. Since we follow a double-entry system of accounts, the total of all the debit and credit balances as appeared in the trial balance remains equal.	Preparation of Trial Balance is also an *automatic process* in the system. You can generate a trial balance from the system anytime.

5. Posting of Adjustment Entries	In this step, the adjustment entries are first passed through the journal, followed by posting in ledger accounts, and finally in the trial balance. Since in most of the cases, we used the accrual basis of accounting to find out the correct value of revenue, expenses, assets, and liabilities accounts, we need to do these adjustment entries. This process is performed at the end of each accounting period.	General Journal is used for posting the adjustment entries in the system. Again, the posting to the ledgers is an automatic process in D365.
6. Adjusted Trial Balance	Considering the above adjustment entries, we create an adjusted trial balance. Adjusted trial balance is a platform to prepare the financial statements of a company.	Trail Balance is *automatically adjusted* because of the posting of the journals in the system. So, no extra process or step is required.
7. Preparation of Financial Statements	Financial statements are the set of statements like Profit & Loss Account, Cash Flow Statement, Balance Sheet. With the help of trial balance, we put all the information into financial statements. Financial statements clearly show the financial health of a firm by depicting its profits or losses.	Financial Statements can be generated in 'Dynamics 365' in *real-time* as soon as the journals are posted in the system. Designing the financial statements is a one-time activity, after that, all the reports can automatically be generated.
8. Post-Closing Entries	All the different nominal accounts of the company are transferred to the Profit & Loss account. With the result of these entries, the balance of all the nominal accounts comes to NIL. The net balance of these entries represents the profit or loss of the company, which is finally accounting. We pass these entries	There is a closing process in 'Dynamics 365' which *automatically generates the closing entries* and carries forward the net balance of nominal accounts to the reserve and surplus account and carries forward all the

	only at the end of the accounting period.	assets & liabilities to the next year.
9. Post-Closing Trial Balance	Post-closing Trial Balance represents the balances of Asset, Liabilities & Capital account. These balances are transferred to the next financial year as an opening balance.	Trail Balance is *automatically adjusted* because of the posting of the journals in the system. So, no extra process or step is required.

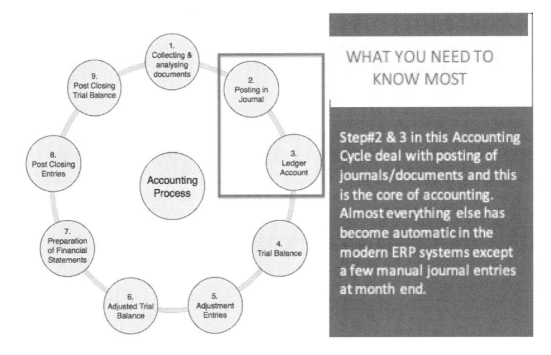

WHAT YOU NEED TO KNOW MOST

Step#2 & 3 in this Accounting Cycle deal with posting of journals/documents and this is the core of accounting. Almost everything else has become automatic in the modern ERP systems except a few manual journal entries at month end.

What is a general ledger account?

A *general ledger account* is an account or record used to sort and store balance sheet and income statement transactions. An account is identified by a numeric code.

We have seen the examples of general ledger accounts in the previous chapter also, following are the example below:

Category	Example
Asset Accounts	Cash, Accounts Receivable, Inventory, furniture & fixture
Liability Accounts	Notes Payable, Accounts Payable, Accrued Expenses Payable, Customer Deposits

Owner's Equity	Share Capital, Reserves
Income Accounts	Sales, Service Fee Revenues
Expense Accounts	Salaries Expense, Rent Expense, Advertising Expense, Interest Expense, and Loss on Disposal of Assets

Chart of Accounts- A listing of a company's general ledger accounts is found in its Chart of Accounts.

Control Accounts & Subsidiary Ledgers

A control account is a summary account in the general ledger. The details that support the balance in the summary account are contained in a subsidiary ledger—a ledger outside of the general ledger.

The purpose of the control account is to keep the general ledger free of details, yet have the correct balance for the financial statements. For example, the Accounts Receivable account in the general ledger could be a control account. If it were a control account, the company would merely update the account with a few amounts, such as total collections for the day, total sales on account for the day, total returns and allowances for the day, etc.

The details on each customer and each transaction would not be recorded in the Accounts Receivable control account in the general ledger. Rather, these details of the accounts receivable activity will be in the Accounts Receivable *Subsidiary Ledger*. This works well because the employees working with the general ledger probably do not need to see the details for every sale or every collection transaction. However, the sales manager and the credit manager will need to know detailed information on individual customers, including whether a customer recently reduced their account balance. The company can provide these individuals with access to the Accounts Receivable Subsidiary Ledger and can keep the general ledger free of a tremendous amount of detail.

Everything that is posted into Sub-ledgers is also posted into General Ledger and they act together to provide progressive levels of detail/summary.

Example: -

See this pictorial presentation of a customers' subledger with its summary account (sub-ledger) 'Accounts receivable' account.

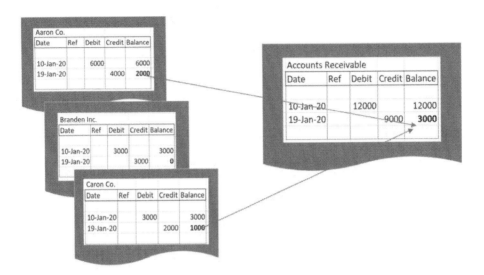

The total debits and credits in Accounts Receivable in the general ledger are reconcilable to the detailed debits and credits in the subsidiary accounts. The balance of $3,000 in the Accounts Receivable control account agrees with the total of the balances in the individual accounts receivable accounts ($2,000 + $0 + $1,000) in the subsidiary ledger. The biggest point here is that when you see the trial balance, only the 'Accounts receivable' account will be there instead of individual customers. It keeps the financials light and shows the level of details required at that level.

How Sub-Ledger posting is different from Ledger posting?

If you remember this cycle, all the transactions are ultimately posted to the ledger (step 3) when the journal is processed.

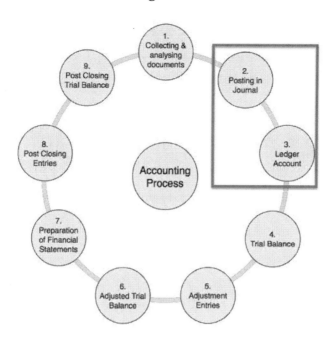

But some ledgers cannot be directly transacted but the transactions must be routed through sub-ledgers. The image below gives a very nice presentation of posting. All the journals are going to the ledger through a subsidiary ledger except a few which are directly hitting the ledger account which simply means there is no need to maintain a sub-ledger for those transactions. It will be crystal clear for you in the next section.

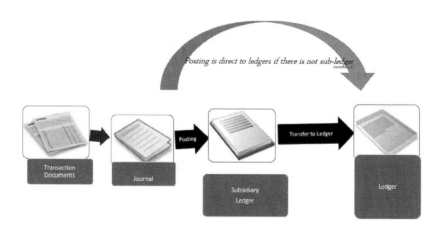

See the below summary of the sub-ledgers which are generally used in any system.

Ledger	Sub-Ledger	Module Name in D365
Fixed Assets Control A/c	Fixed Assets Master	Fixed Assets Module
Cash & Bank A/c	Bank Master	Bank Module
Accounts Receivable A/c	Customer Master	Accounts Receivable Module
Inventory A/c	Product Master	Product Information Management
Accounts Payable	Vendor Master	Accounts Payable Module

💡 Just remember that whatever ledger is identified as control account or sub-ledger, those accounts cannot be directly used in the journals. ERP will take care of everything once the chart of accounts is properly configured, so you don't need to worry too much.

Don't miss the next section to understand how the actual posting happens in the ledger and sub-ledger and how it impacts financial statements. It will make the concept clearer for you.

Examples of Ledger Posting and impact on Financials.

Let's take the same example from chapter 2 to understand the concept of posting.

Transaction#1 Introduction of Share Capital
When the investment money is deposited into the bank by the investors-

Debit/ Credit	Type of Account	Account details	Class of Account	Rule#	Description

| Dr. | Asset | Bank A/c | Real | #1 | Debit what comes in / Debit the increase in the assets |
| Cr | Capital | Share Capital A/ | Personal | #1 | Credit the increase in the liability |

Ledger Posting→

- ✓ *The bank account is debited*

- ✓ *'Share Capital A/c' is credited.*

Bank A/c

Date	Description	Debit ($)	Credit ($)	Balance($)
05-Jan-21	Deposit (share capital	100,000.00		100,000.00

Share Capital A/c

Date	Description	Debit ($)	Credit ($)	Balance($)
05-Jan-21	Deposit (share capital)		100,000.00	(100,000.00)

Impact on Financial Statement →

Only the balance sheet will be impacted here as there is no account hitting the profit & loss accounts.

Balance Sheet as on ....

	Amount ($)
Cash & cash equivalents	100,000.00
Trade & other receivables	-
Intangible Assets	-
Other non-current assets	-
Total Assets (A)	100,000.00
Trade & other payables	-
Other current liabilities	-
Borrowings	-
Total Liabilities (B)	-
Net Assets (A-B)	100,000.00
Share Capital	100,000.00
Retained earnings & reserves	
Total Equity (C)	100,000.00

> 1. See the color coding above to match the source and destination of ledger balance.
>
> 2. This balance sheet is shown above in vertical format which is a common practice now. Old template used to show 'Liabilities' on the left column and 'Assets' on the right column.

Transaction#2 *Purchase of Fixed Assets*

ABC furniture needs some assets (e.g., computers, machines) in the back office and factory to manufacture the furniture or keeping the books. So, the purchasing department buys some assets from the vendors. (We are buying machinery for factory and furniture for office in this example)

Debit/ Credit	Type of Account	Account details	Class of Account	Rule#	Description
Dr.	Asset	Machinery & Equipment A/c	Real	#1	Debit the increase in

					the assets
Cr.	Liability	Dell computers (AP) A/c	Personal	#1	Credit the increase in the liability

Debit/ Credit	Type of Account	Account details	Class of Account	Rule#	Description
Dr.	Asset	Furniture & Fixture A/c	Real	#1	Debit the increase in the assets
Cr.	Liability	Ikea Furniture	Personal	#1	Credit the increase in the liability

Ledger Posting→

Office Equipment A/c

Date	Description	Debit ($)	Credit ($)	Balance($)
10-Jan-21	Laptop from Dell	20,000.00		20,000.00

Furniture & Fixture A/c

Date	Description	Debit ($)	Credit ($)	Balance($)
11-Jan-21	Office desk from Ikea	8,000.00		8,000.00

Dell Computers A/c

Date	Description	Debit ($)	Credit ($)	Balance($)
10-Jan-21	Laptop from Dell		20,000.00	(20,000.00)

Ikea Furniture

Date	Description	Debit ($)	Credit ($)	Balance($)
11-Jan-21	Office desk from Ikea		8,000.00	(8,000.00)

Accounts Payable (non Trade) A/c

Date	Description	Debit ($)	Credit ($)	Balance($)
10-Jan-21	Laptop from Dell		20,000.00	(20,000.00)
11-Jan-21	Office desk from Ikea		8,000.00	(28,000.00)

Impact on Financial Statements→

Balance Sheet as on ……..

	Amount ($)
Cash & cash equivalents	100,000.00
Trade & other receivables	
Intangible Assets	
Other non-current assets	28,000.00
Total Assets (A)	128,000.00
Trade & other payables	
Other current liabilities	28,000.00
Borrowings	
Total Liabilities (B)	28,000.00
Net Assets (A-B)	100,000.00
Share Capital	100,000.00
Retained earnings & reserves	
Total Equity (C)	100,000.00

> ⚙️ 'Dell Computers' and 'Ikea Furniture' both are sub-ledger in this example, and you will see that the total of these appear in the balance sheet under 'Accounts Payable Non- Trade' ledger which is mapped in the heading 'Other Non-Current Assets'. In other words, individual vendors are under a control account which goes into financials.

Transaction#3 Payment to the non-trade suppliers.

The invoice of the Ikea furniture must be paid immediately, the amount is transferred from the bank to the supplier.

Debit/ Credit	Type of Account	Account details	Class of Account	Rule#	Description
Dr.	Liability	Ikea Furniture	Personal	#1	Debit the decrease in the liability
Cr.	Asset	Bank A/c	Real	#1	Credit the decrease in the asset

Ledger Posting→

Ikea Furniture

Date	Description	Debit ($)	Credit ($)	Balance($)
11-Jan-21	Office desk from Ikea		8,000.00	(8,000.00)
11-Jan-21	Invoice paid	8000		-

Accounts Payable (non Trade) A/c

Date	Description	Debit ($)	Credit ($)	Balance($)
10-Jan-21	Laptop from Dell		20,000.00	(20,000.00)
11-Jan-21	Office desk from Ikea		8,000.00	(28,000.00)
11-Jan-21	Invoice paid	8000		(20,000.00)

Bank A/c

Date	Description	Debit ($)	Credit ($)	Balance($)
05-Jan-21	Deposit (share capital	100,000.00		100,000.00
11-Jan-21	Invoice paid (Ikea)		8,000.00	92,000.00

Impact on Financial Statements→

Balance Sheet as on ....

	Amount ($)
Cash & cash equivalents	92,000.00
Trade & other receivables	-
Intangible Assets	-
Other non-current assets	28,000.00
Total Assets (A)	120,000.00
Trade & other payables	-
Other current liabilities	20,000.00
Borrowings	-
Total Liabilities (B)	20,000.00
Net Assets (A-B)	100,000.00
Share Capital	100,000.00
Retained earnings & reserv	-
Total Equity (C)	100,000.00

Transaction#4 Purchase of Stock

Now, there is some furniture which is bought for trading purposes since ABC Furniture Ltd. is a furniture company.

Debit/ Credit	Type of Account	Account details	Class of Account	Rule#	Description
Dr.	Asset	Inventory A/c	Real	#1	Debit the increase in the assets
Cr.	Liability	Vendor (AP) A/c	Personal	#1	Credit the increase in the liability

✓ The accounting treatment is almost the same in this case also except the nature of the real account is 'Inventory account' instead of 'Furniture & Fixture Account.'

Ledger Posting→

Inventory A/c

Date	Description	Debit ($)	Credit ($)	Balance($)
20-Jan-21	Bought stock (100 chairs @$90 each)	9,000.00		9,000.00

ABC Furniture Ltd   (SUB Ledger)

Date	Description	Debit ($)	Credit ($)	Balance($)
20-Jan-21	Bought stock (100 chairs @$90 each)		9,000.00	(9,000.00)

Accounts Payable (Trade) A/c

Date	Description	Debit ($)	Credit ($)	Balance($)
20-Jan-21	Bought stock (100 chairs @$90 each)		9,000.00	(9,000.00)

Impact on Financial Statements→

Balance Sheet as on ....

	Amount ($)
Cash & cash equivalents	92,000.00
Trade & other receivables	-
Intangible Assets	-
Inventory	9,000.00
Other non-current assets	28,000.00
Total Assets (A)	129,000.00
Trade & other payables	9,000.00
Other current liabilities	20,000.00
Borrowings	-
Total Liabilities (B)	29,000.00
Net Assets (A-B)	100,000.00
Share Capital	100,000.00
Retained earnings & reserv	-
Total Equity (C)	100,000.00

Transaction#5 Freight on Stock paid

Payment made to the transporter who brought the furniture to the warehouse-

Debit/ Credit	Type of Account	Account details	Class of Account	Rule#	Description
Dr.	Expense	Freight A/c	Nominal	#2	Debit all the expenses/losses
Cr.	Asset	Bank A/c	Real	#1	Credit the decrease in the assets

- ✓ Freight is a nominal account, so going by rule#2, it is debited here and

- ✓ Bank which is a real account, and is decreasing, so it is credited here.

-☼- Instead of debiting the 'freight A/c', we could also debit the 'Stock A/c' since it is adding the value of the stock. But don't worry about this treatment yet. Just focus on the basic accounting for now.

Ledger Posting→

Freight Charges A/c

Date	Description	Debit ($)	Credit ($)	Balance($)
21-Jan-21	Freight paid	1,500.00		1,500.00

Bank A/c

Date	Description	Debit ($)	Credit ($)	Balance($)
05-Jan-21	Deposit (share capital	100,000.00		100,000.00
11-Jan-21	Invoice paid (Ikea)		8,000.00	92,000.00
21-Jan-21	Freight paid		1,500.00	90,500.00

Impact on Financial Statements→

Balance Sheet as on ....

	Amount ($)
Cash & cash equivalents	90,500.00
Trade & other receivables	-
Intangible Assets	-
Inventory	9,000.00
Other non-current assets	28,000.00
Total Assets (A)	127,500.00
Trade & other payables	9,000.00
Other current liabilities	20,000.00
Borrowings	-
Total Liabilities (B)	29,000.00
Net Assets (A-B)	98,500.00
Share Capital	100,000.00
Retained earnings & reserv	(1,500.00)
Total Equity (C)	98,500.00

Profit & Loss for the period....

	Amount ($)
Revenue (Sales)	-
Operating Expenses	1,500.00
EBITDA before non-operating e	(1,500.00)
Non-operating expense	-
Depreciation	
Profit before tax	(1,500.00)
Taxation	-
Net profit/loss (NPAT)	(1,500.00)

Transaction#6 Sale of stock to a customer

A customer orders 10 chairs. Following is the accounting entry for a sale transaction-

Debit/ Credit	Type of Account	Account details	Class of Account	Rule#	Description
Dr.	Asset	Customer (AR) A/c	Personal	#1	Debit the increase in the assets
Cr.	Revenue	Sales A/c	Nominal	#2	Credit all income/profit
Dr.	Expense	Cost of Goods Sold	Nominal	#2	Debit all the expenses/losses
Cr.	Asset	Inventory A/c	Real	#1	Credit the decrease in the assets

This transaction is a bit complex for beginners but let's try to decode this also in two parts. Part-I

✓ The customer account will be debited because he is an asset since we need to recover money from him after we sold goods to him on credit.

✓ sales account (revenue) account will be credited (rule#2) since it is the income of the business.

Part-II

✓ The inventory account will be credited with the cost amount (rule#1) since it is a real account and it is decreasing

✓ The Debit side of this transaction will be 'Cost of Goods Sold (COGS)' which means that we are converting that inventory into our cost now and all expenses are debited.

Ledger Posting→

Sales A/c

Date	Description	Debit ($)	Credit ($)	Balance($)
28-Jan-21	Sold 30 chairs @$150 each		4500	(4,500.00)

University of Sydney (Customer SUB Ledger)

Date	Description	Debit ($)	Credit ($)	Balance($)
28-Jan-21	Sold 30 chairs @$150 each	4,500.00		4,500.00

Accounts Receivable A/c

Date	Description	Debit ($)	Credit ($)	Balance($)
28-Jan-21	Sold 30 chairs @$150 each	4,500.00		4,500.00

Cost of Goods Sold A/c

Date	Description	Debit ($)	Credit ($)	Balance($)
28-Jan-21	COGS for 30 chairs sold @$90 each	2,700.00		2,700.00

Inventory A/c

Date	Description	Debit ($)	Credit ($)	Balance($)
20-Jan-21	Bought stock (100 chairs @$90 each)	9,000.00		9,000.00
28-Jan-21	COGS for 30 chairs sold @$90 each		2,700.00	6,300.00

Financial Accounting with Microsoft D365 ERP

Impact on Financial Statements→

Balance Sheet as on ....

	Amount ($)
Cash & cash equivalents	90,500.00
Trade & other receivables	4,500.00
Intangible Assets	-
Inventory	6,300.00
Other non-current assets	28,000.00
Total Assets (A)	129,300.00
Trade & other payables	9,000.00
Other current liabilities	20,000.00
Borrowings	-
Total Liabilities (B)	29,000.00
Net Assets (A-B)	100,300.00
Share Capital	100,000.00
Retained earnings & reserv	300.00
Total Equity (C)	100,300.00

Profit & Loss for the period....

	Amount ($)
Revenue (Sales)	4,500.00
Operating Expenses	4,200.00
EBITDA before non-operating e	300.00
Non-operating expense	-
Depreciation	
Profit before tax	300.00
Taxation	-
Net profit/loss (NPAT)	300.00

💡 Operating expenses heading is the total of 'Cost of Goods sold' and 'Freight Expenses'

Transaction#7 Collection from the customer (Partial)

The customer makes the payment for the purchase he made from ABC

Debit/ Credit	Type of Account	Account details	Class of Account	Rule#	Description
Dr.	Asset	Bank A/c	Real	#1	Debit the increase in the assets
Cr.	Asset	Customer (AR) A/c	Personal	#1	Credit the decrease in the assets

✓ Since the bank balance will increase, it will be debited and

✓ the customer account (an asset) is decreasing, and it will be credited as per rule#1.

Ledger Posting→

Bank A/c

Date	Description	Debit ($)	Credit ($)	Balance($)
05-Jan-21	Deposit (share capital)	100,000.00		100,000.00
11-Jan-21	Invoice paid (Ikea)		8,000.00	92,000.00
21-Jan-21	Freight paid		1,500.00	90,500.00
########	Part payment UTS	2000		92,500.00

University of Sydney   (Customer SUB Ledger)

Date	Description	Debit ($)	Credit ($)	Balance($)
28-Jan-21	Sold 30 chairs @$150 each	4,500.00		4,500.00
########	Part payment		2,000.00	2,500.00

Accounts Receivable A/c

Date	Description	Debit ($)	Credit ($)	Balance($)
28-Jan-21	Sold 30 chairs @$150 each	4,500.00		4,500.00
########	Part payment		2,000.00	2,500.00

Impact on Financial Statements→

Balance Sheet as on ....

	Amount ($)
Cash & cash equivalents	92,500.00
Trade & other receivables	2,500.00
Intangible Assets	-
Inventory	6,300.00
Other non-current assets	28,000.00
Total Assets (A)	129,300.00
Trade & other payables	9,000.00
Other current liabilities	20,000.00
Borrowings	-
Total Liabilities (B)	29,000.00
Net Assets (A-B)	100,300.00
Share Capital	100,000.00
Retained earnings & reserve	300.00
Total Equity (C)	100,300.00

Profit & Loss for the period....

	Amount ($)
Revenue (Sales)	4,500.00
Operating Expenses	4,200.00
EBITDA before non-operating e	300.00
Non-operating expense	-
Depreciation	
Profit before tax	300.00
Taxation	-
Net profit/loss (NPAT)	300.00

As you can see any accounting entry has two aspects and it is posted in some accounts ultimately. The system takes care of all these steps, and it makes sure that the trial balance is also the balanced end of the day. Gone are the days when it was all manual and the accountants used to spend hundreds of hours just to reconcile the trial balance

Visit our website www.satyakejriwal.com for more blogs and videos on accounting.

4

Best Practices- Chart of Accounts

A chart of accounts is a listing of the names of the accounts that a company has identified and made available for recording transactions in its general ledger. A company has the flexibility to tailor its chart of accounts to best suit its needs, including adding accounts as needed.

Within the chart of accounts, you will find that the accounts are typically listed in the following order:

Balance Sheet Accounts	Assets
	Liabilities
	Owner's Equity (Share Capital/Surplus)
Income Statement Accounts	Revenue Accounts
	Expense Accounts

Within the categories of revenues and expenses, accounts might be further organized by business function (such as producing, selling, administrative, financing) and/or by company divisions, product lines, etc.

A company's organization chart can serve as the outline for its accounting chart of accounts. For example, if a company divides its business into ten departments (production, marketing, human resources, etc.), each department will likely be accountable for its expenses (salaries, supplies, phone, etc.). Each department will have its phone expense account, its own salaries expense, etc.

A chart of accounts will likely be as large and as complex as the company itself. An international corporation with several divisions may need thousands of accounts, whereas a small local retailer may need as few as one hundred accounts.

Sample Chart of Accounts in 'Dynamics 365FO'

> D365 gives the capability to define financial dimensions in the system which helps to keep a Chart of Accounts slim e.g., divisions and departments can be defined as financial dimensions, and you don't need to set up ledger for each division and department. So, if there is only one account defined for salary expense in the system, you can still report on salary for each department and division.

Each account in the chart of accounts is typically assigned a name and a unique number by which it can be identified. (Software for some small businesses may not require account numbers.) Account numbers are often five or more digits in length with each digit representing a division of the company, the department, the type of account, etc.

As you will see, the first digit might signify if the account is an asset, liability, etc. For example, if the first digit is a "1" it is an asset. If the first digit is a "5" it is an operating expense. Similarly, the next digit in the code can represent subgrouping e.g. '0' after 1 in the following accounts represent cash accounts

    10100 Cash - Regular Checking
    10200 Cash - Payroll Checking

The following is a partial listing of a sample chart of accounts.

Current Assets (account numbers 10000 - 16999)

    10100 Cash - Regular Checking
    10200 Cash - Payroll Checking
    10600 Petty Cash Fund
    12100 Accounts Receivable
    12500 Allowance for Doubtful Accounts
    13100 Inventory
    14100 Supplies

15300 Prepaid Insurance

Property, Plant, and Equipment (account numbers 17000 - 18999)

17000 Land
17100 Buildings
17300 Equipment
17800 Vehicles
18100 Accumulated Depreciation - Buildings
18300 Accumulated Depreciation - Equipment
18800 Accumulated Depreciation - Vehicles

Current Liabilities (account numbers 20010 - 24999)

20110 Notes Payable - Credit Line #1
20210 Notes Payable - Credit Line #2
21000 Accounts Payable
22100 Wages Payable
23100 Interest Payable
24500 Unearned Revenues

Long-term Liabilities (account numbers 25000 - 26999)

25100 Mortgage Loan Payable
25600 Bonds Payable
25650 Discount on Bonds Payable

Stockholders' Equity (account numbers 27000 - 29999)

27100 Common Stock, No Par
27500 Retained Earnings
29500 Treasury Stock

Operating Revenues (account numbers 30000 - 39999)

31010 Sales - Division #1, Product Line 010
31022 Sales - Division #1, Product Line 022

32016 Sales - Division #2, Product Line 015
33110 Sales - Division #3, Product Line 110

Cost of Goods Sold (account numbers 40000 - 49999)

41010 COGS - Division #1, Product Line 010
41022 COGS - Division #1, Product Line 022
42016 COGS - Division #2, Product Line 015
43110 COGS - Division #3, Product Line 110

Marketing Expenses (account numbers 50000 - 50999)

50100 Marketing Dept. Salaries
50150 Marketing Dept. Payroll Taxes
50200 Marketing Dept. Supplies
50600 Marketing Dept. Telephone

Other (account numbers 90000 - 99999)

91800 Gain on Sale of Assets
96100 Loss on Sale of Assets

Best Practices for Designing a Chart of Accounts in D365
Shared CoA or company-wise CoA

This is one of the major decisions in a multiple legal entity scenario. Some legal entities will need this own chart of accounts for their own legal = or local language requirements e.g., China, Japan, France. So, it is always recommended not to mix their accounts with a common chart of accounts.

Numbering of ledger and scalability

The numbering should allow for further extension and should be scalable. Make sure there are enough gaps in each grouping so that more accounts can be inserted in the future. In the example below, you can see there is enough gap between the two accounts. If there are more sales accounts that need to be inserted after 31022, it is easily achievable.

> 31010 Sales - Division #1, Product Line 010
> 31022 Sales - Division #1, Product Line 022
> 32016 Sales - Division #2, Product Line 015
> 33110 Sales - Division #3, Product Line 110

Another tip on the number is not to keep the numbering too long or too short.

Defining control accounts and keeping them closed for direct entry

I always recommend the clients keep their chart of accounts slim and leverage the sub-leger accounts. Most of the organizations who are still running on 20-30 years old home-grown systems have 2000 ledgers long chart of accounts ⬚ and there is a good reason behind that. Those systems were not future-proof and were not scalable.

> o Keep individual bank accounts out of the chart of accounts and just create a few summary accounts e.g.
>
>> 10100 Cash - Regular Checking
>> 10200 Cash - Payroll Checking
>
> o Keep individual customers and vendors in AR and AP modules and create AR/AP control accounts e.g.
>
>> 12100 Accounts Receivable
>>
>> 21000 Accounts Payable
>> 22100 Wages Payable
>> 23100 Interest Payable
>
> o Keep individual fixed assets in the Fixed Asset Module and create fixed assets control accounts e.g.

17000 Land

17100 Buildings

17300 Equipment

17800 Vehicles

The same goes with inventory accounts, sales accounts, purchases accounts. In modern ERP systems, there is a special module for each sub-ledger and all the details are limited to those modules.

Ledgers should be optimized with 'Financial dimensions'.

Another way to keep the chart of accounts short is financial dimensions in D365. Ledger and dimensions together make a matrix, and we can avoid creating ledger accounts per department, per business unit, per geographical region, etc.

💡 *What is the purpose of subsidiary ledgers?*

In Accounting, there are two types of Ledgers, the General Ledger (Book of final entry) and Subsidiary (Sub) Ledgers. The Accounts for the General Ledger come from the Chart of Accounts. A subsidiary ledger is a group of accounts with a common characteristic, e.g., all are customer accounts, that is, all are accounts receivable. The subsidiary ledger facilitates the recording process by freeing the general ledger from the details of individual balances.

See the below summary of the sub-ledgers which are generally used in any system.

Ledger	Sub-Ledger	Modules in D365FO
Fixed Assets Control A/c	Fixed Assets Master	Fixed Assets Module
Cash & Bank A/c	Bank Master	Bank Module
Accounts Receivable A/c	Customer Master	Accounts Receivable Module
Inventory A/c	Product Master	Product Information Management
Accounts Payable	Vendor Master	Accounts Payable Module

For example – just create a 'Travelling Expense' account in the chart of accounts and create 3 financial dimensions e.g. department, business unit, and region. There might be 4 departments, 6 business units, and 2 regions. Ideally, there should be a minimum of 12 types of travelling expense account (or maybe more) to cover each department, unit, and region but we can just create one account and use the dimensions to tag the account during voucher entry.

5

IFRS and International Accounting Standards

International Financial Reporting Standards (IFRS), issued by the International Accounting Standards Board (IASB), is rapidly becoming a benchmark for the accounting world. IFRS are like any other accounting standards prevailing in many countries. Just like any other accounting standard, IFRS also helps in the standardization of accounting policies and methods of valuation and recognition of assets, liabilities, income, and expenditures, etc.

These are internationally designed accounting standards prepared after detailed research to set principles for the preparation of financial statements of any organization. These principles help in the standardization of financial statements and help a user of financial statements in understanding the true and fair picture of the financial affairs of the concerned organization.

If financial statements are prepared using the same principles worldwide, it is easy to compare, merge or amalgamate the financial statements of different

Why IAS and IFRS co-exist?

Accounting standard issued by IASC Board are known as IAS and w.e.f. April 1, 2001 Accounting standard issued by IASB is known as IFRS. IASB, according to changed global accounting practices, is making changes in IAS also and new updated IAS are being replaced by new series of IFRS.

In simple words, IAS are now transitioning to IFRS series.

organizations situated in a geographically different location or doing business in different business segments. Thus, helps the user of financial statements in taking a useful decision like investment/disinvestment, etc.

About the International Accounting Standards Board (Board)

The International Accounting Standards Board (IASB) is an independent, private-sector body that develops and approves International Financial Reporting Standards (IFRSs). The IASB operates under the oversight of the IFRS Foundation. The IASB was formed in 2001 to replace the International Accounting Standards Committee (IASC). A full history of the IASB and the IASC going back to 1973 is available on the IASB website.
Currently, the IASB has 14 members.
The IASB's role

Under the IFRS Foundation Constitution, the IASB has complete responsibility for all technical matters of the IFRS Foundation including:

- full discretion in developing and pursuing its technical agenda, subject to certain consultation requirements with the Trustees and the public
- the preparation and issuing of IFRSs (other than Interpretations) and exposure drafts, following the due process stipulated in the Constitution
- the approval and issuing of Interpretations developed by the IFRS Interpretations Committee.

Summary of IAS/IFRS/IFRIC

Following is a list of IFRS, IAS which are currently applicable. We have covered some of them in this book and the rest of them will be in the next book of this series. Secondly, some of the IAS/IFRS doesn't warrant a **detailed** explanation, so we have given just the **reference** of those. Moreover, not all the requirements of IFRS can be system-driven.

International Accounting Standards				
#	Name	Issued	Covered In	Detailed or Reference
IAS 1	Presentation of Financial Statements	2007*	Book-1	Detailed
IAS 2	Inventories	2005*	Book-1	Detailed
IAS 7	Statement of Cash Flows	1992	Book-1	Detailed
IAS 8	Accounting Policies, Changes in Accounting Estimates and Errors	2003	Book-1	Detailed
IAS 10	Events After the Reporting Period	2003	Book-1	Detailed
IAS 12	Income Taxes	1996*	Book-2	
IAS 16	Property, Plant and Equipment	2003*	Book-1	Detailed
IAS 19	Employee Benefits (2011)	2011*	Book-2	
IAS 20	Accounting for Government Grants and Disclosure of Government Assistance	1983	Book-2	
IAS 21	The Effects of Changes in Foreign Exchange Rates	2003*	Book-1	Detailed
IAS 23	Borrowing Costs	2007*	Book-2	
IAS 24	Related Party Disclosures	2009*	Book-2	
IAS 26	Accounting and Reporting by Retirement Benefit Plans	1987	Book-2	
IAS 27	Separate Financial Statements (2011)	2011	Book-2	
IAS 28	Investments in Associates and Joint Ventures (2011)	2011	Book-2	
IAS 29	Financial Reporting in Hyperinflationary Economies	1989	Book-1	Reference
IAS 32	Financial Instruments: Presentation	2003*	Book-2	
IAS 33	Earnings Per Share	2003*	Book-2	
IAS 34	Interim Financial Reporting	1998	Book-1	Detailed
IAS 36	Impairment of Assets	2004*	Book-1	Detailed
IAS 37	Provisions, Contingent Liabilities and Contingent Assets	1998	Book-1	Detailed
IAS 38	Intangible Assets	2004*	Book-1	Reference

IAS 40	Investment Property	2003*	Book-1	Detailed
IAS 41	Agriculture	2001	Book-2	

IFRS				
#	Name	Issued	Covered In	Detailed or Reference
IFRS 1	First-time Adoption of International Financial Reporting Standards	2008*	Book-2	
IFRS 2	Share-based Payment	2004	Book-2	
IFRS 3	Business Combinations	2008*	Book-1	Detailed
IFRS 5	Non-current Assets Held for Sale and Discontinued Operations	2004	Book-1	Reference
IFRS 6	Exploration for and Evaluation of Mineral Resources	2004	Book-2	
IFRS 7	Financial Instruments: Disclosures	2005	Book-2	Reference
IFRS 8	Operating Segments	2006	Book-1	Detailed
IFRS 9	Financial Instruments	2014*	Book-2	
IFRS 10	Consolidated Financial Statements	2011	Book-1	Detailed
IFRS 11	Joint Arrangements	2011	Book-1	Detailed
IFRS 12	Disclosure of Interests in Other Entities	2011	Book-2	
IFRS 13	Fair Value Measurement	2011	Book-1	Reference
IFRS 14	Regulatory Deferral Accounts	2014	Book-2	
IFRS 15	Revenue from Contracts with Customers	2014	Book-2	
IFRS 16	Leases	2016	Book-2	
IFRS 17	Insurance Contracts	2017	Book-2	

Note- Highlighted is the standards that are covered in this book.

How many countries follow IFRS now?

Approximately <u>120 nations have adopted IFRS completely or partially</u> for reporting of financials by their domestic listed/non listed companies. Maximum countries have adopted IFRS as promulgated by IASB. Companies situated in these countries, while publishing their financials, include a statement acknowledging such <u>adoption/conformity in its audit report.</u>

Few countries though adopted IFRS but not same as promulgated by IASB. Their respective statutory accounting governing bodies/boards have <u>adopted IFRS by conversing their existing standard into new set of accounting standards which is mostly at par with IFRS,</u> with certain carve-outs. Reason of these carve-outs is mainly different socioeconomic, commercia and statutory behavior/practices existing in those countries.

Presentation and Disclosure relation IFRS/IAS

Many of the IFRS and IAS prescribes how to present the financial statements. IAS 1 and IFRS 10 are the most important ones in this list and others are somehow related. Let's see the main provisions of these standards and how D365 helps in meeting the requirements:

#	Name	Issued
IAS 1	***Presentation of Financial Statements***	2007*
IAS 7	*Statement of Cash Flows*	1992
IAS 8	*Accounting Policies, Changes in Accounting Estimates and Errors*	2003
IAS 10	*Events After the Reporting Period*	2003
IAS 29	*Financial Reporting in Hyperinflationary Economies*	1989
IAS 34	*Interim Financial Reporting*	1998
IFRS 3	*Business Combinations*	2008*
IFRS 8	*Operating Segments*	2006
IFRS 10	***Consolidated Financial Statements***	2011
IFRS 11	*Joint Arrangements*	2011

IAS 1 — Presentation of Financial Statements

IAS 1 Presentation of Financial Statements sets out the overall requirements for financial statements, including how they should be structured, the minimum requirements for their content, and overriding concepts such as going concern, the accrual basis of accounting, and the current/non-current distinction.

Components of financial statements

IAS-A prescribes that a complete set of financial statements should include the following statements

- statement of financial position (balance sheet) at the end of the period
- a statement of profit or loss and other comprehensive income for the period (presented as a single statement, or by presenting the profit or loss section in a separate statement of profit or loss, immediately followed by a statement presenting comprehensive income beginning with profit or loss)
- a statement of changes in equity for the period
- a statement of cash flows for the period
- notes, comprising a summary of significant accounting policies and other explanatory notes
- comparative information prescribed by the standard.

An entity may use titles for statements other than those stated above. All financial statements are required to be presented with equal prominence.

The purpose and structure of financial statements can be understood with the following flow chart

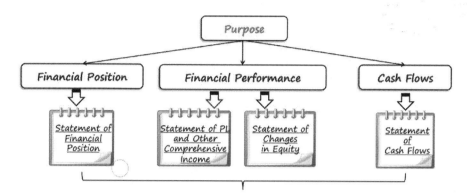

NOTES TO THE FINANCIAL STATEMENTS

Formats of Financial Statements

Though IAS-1 prescribes the purpose and structure of financial statements it does not prescribe any standard format for the presentation of financial statements. IAS-1 also prescribes the line items which is included while presenting any financial statement

(A) Balance Sheet/ Statement of financial statement. The line items to be included on the face of the statement of financial position (Balance Sheet) are:

(a)	property, plant, and equipment
(b)	investment property
(c)	intangible assets
(d)	financial assets (excluding amounts shown under (e), (h), and (i))
(e)	investments accounted for using the equity method
(f)	biological assets
(g)	Inventories

(h)	trade and other receivables
(i)	cash and cash equivalents
(j)	assets held for sale
(k)	trade and other payables
(l)	Provisions
(m)	financial liabilities (excluding amounts shown under (k) and (l))
(n)	current tax liabilities and current tax assets, as defined in IAS 12
(o)	deferred tax liabilities and deferred tax assets, as defined in IAS 12
(p)	liabilities included in disposal groups
(q)	non-controlling interests, presented within equity
(r)	issued capital and reserves attributable to owners of the parent.

Additional line items, headings, and subtotals may be needed to fairly present the entity's financial position.

(B) Statement of profit or loss and other comprehensive income

Concepts of profit or loss and comprehensive income: Profit or loss is defined as "the total of income less expenses, excluding the components of other comprehensive income".

Other comprehensive income is defined as comprising "items of income and expense (including reclassification adjustments) that are not recognized in profit or loss as required or permitted by other IFRSs".

Total comprehensive income is defined as "the change in equity during a period resulting from transactions and other events, other than those changes resulting from transactions with owners in their capacity as owners".

Comprehensive income for the period	=	Profit or loss	+	Other comprehensive income

Choice in presentation and basic requirements: An entity has a choice of presenting:

- a single statement of profit or loss and other comprehensive income, with profit or loss and other comprehensive income presented in two sections, or
- two statements:
  - ✓ a separate statement of profit or loss
  - ✓ a statement of comprehensive income, immediately following the statement of profit or loss and beginning with profit or loss

The statement(s) must present:

- ➤ profit or loss
- ➤ total other comprehensive income
- ➤ comprehensive income for the period
- ➤ an allocation of profit or loss and comprehensive income for the period between non-controlling interests and owners of the parent.

Following minimum Line, Items must be presented in the profit or loss section or statement

- ➤ Revenue
- ➤ gains and losses from the derecognition of financial assets measured at amortized cost
- ➤ finance costs
- ➤ share of the profit or loss of associates and joint ventures accounted for using the equity method
- ➤ certain gains or losses associated with the reclassification of financial assets
- ➤ tax expense
- ➤ a single amount for the total of discontinued items

> ➤ Expenses recognized in profit or loss should be analyzed either by nature (raw materials, staffing costs, depreciation, etc.) or by function (cost of sales, selling, administrative, etc.).

Other comprehensive income section

The other comprehensive income section is required to present line items that are classified by their nature and grouped between those items that will or will not be reclassified to profit and loss in subsequent periods.

Other requirements

Additional line items may be needed to fairly present the entity's results of operations

Items cannot be presented as 'extraordinary items in the financial statements or in the notes.

(C) Statement of cash flows

There is a separate standard "IAS 7 Statement of Cash Flows" which explains the requirements of cash flow presentations in detail.

(D) Statement of changes in equity

IAS 1 requires an entity to present a separate statement of changes in equity. The statement must show:

> ➤ total comprehensive income for the period, showing separately amounts attributable to owners of the parent and non-controlling interests
> ➤ the effects of any retrospective application of accounting policies or restatements made under IAS 8, separately for each component of other comprehensive income
> ➤ reconciliations between the carrying amounts at the beginning and the end of the period for each component of equity, separately disclosing:
>   - profit or loss
>   - other comprehensive income

- transactions with owners, showing separately contributions by and distributions to owners and changes in ownership interests in subsidiaries that do not result in a loss of control

(E) Notes to the financial statements

The notes must:

- ➢ present information about the basis of preparation of the financial statements and the specific accounting policies used
- ➢ disclose any information required by IFRSs that is not presented elsewhere in the financial statements and
- ➢ provide additional information that is not presented elsewhere in the financial statements but is relevant to an understanding of any of them
- ➢ Notes are presented systematically and cross-referenced from the face of the financial statements to the relevant note.

IAS 1 suggests that the notes should normally be presented in the following order:

- ➢ a statement of compliance with IFRSs
- ➢ a summary of significant accounting policies applied, including:
  - ❖ the measurement basis (or bases) used in preparing the financial statements
  - ❖ the other accounting policies used that are relevant to an understanding of the financial statements
- ➢ supporting information for items presented on the face of the statement of financial position (balance sheet), statement(s) of profit or loss and other comprehensive income, statement of changes in equity, and statement of cash flows, in the order in which each statement and each line item is presented
- ➢ other disclosures, including:
- ➢ contingent liabilities (refer IAS 37) and unrecognized contractual commitments
- ➢ non-financial disclosures, such as the entity's financial risk management objectives and policies (refer IFRS 7 Financial Instruments: Disclosures)

IAS 7 – Statement of Cash Flow

Objectives:

The objective of IAS 7 is to present the information about the historical changes in cash and cash equivalents of an entity by means of a statement of cash flows, which classifies cash flows during the period according to operating, investing, and financing activities.

Cash Flow Statement Presentation:

Key principles specified by IAS 7 for the preparation of a statement of cash flows are as follows:

❖ Cash flow statement presents the movement in opening and closing cash & cash equivalent because of operating, investing, and financing activities

❖ operating activities are the main revenue-producing activities of the entity that are not investing or financing activities, so operating cash flows include cash received from customers and cash paid to suppliers and employees

❖ investing activities are the acquisition and disposal of long-term assets and other investments that are not considered to be cash equivalents

❖ financing activities are activities that alter the equity capital and borrowing structure of the entity

❖ interest and dividends received and paid may be classified as operating, investing, or financing cash flows, if they are classified consistently from period to period

❖ cash flows arising from taxes on income are normally classified as operating unless they can be specifically identified with financing or investing activities

IAS 8 - Accounting Policies, Changes in Accounting Estimates and Errors

As the name suggests IAS 8 Accounting Policies, Changes in Accounting Estimates and Errors is applied in selecting and applying accounting policies, accounting for changes in estimates, and reflecting corrections of prior period errors.

Accounting Policies: Accounting policies are the specific principles, bases, conventions, rules, and practises applied by an entity in preparing and presenting financial statements.

Change in Accounting Estimates: A change in accounting estimate is an adjustment of the carrying amount of an asset or liability, or related expense, resulting from reassessing the expected future benefits and obligations associated with that asset or liability.

Materiality: Information is material when omitting, misstating, or obscuring it could reasonably be expected to influence decisions that the primary users of general-purpose financial statements make based on those financial statements, which provide financial information about a specific reporting entity.

What are prior period Errors: Prior period errors are omissions from, and misstatements in, an entity's financial statements for one or more prior periods arising from a failure to use, or misuse of, reliable information that was available and could reasonably be expected to have been obtained and taken into account in preparing those statements. Such errors result from mathematical mistakes, mistakes in applying accounting policies, oversights or misinterpretations of facts, and fraud.

How to select Accounting Policies: For any transaction or item if there is any specific Standard or Interpretation then the accounting policy or policies applied to that item must be determined by applying the Standard or Interpretation and considering any relevant Implementation Guidance issued by the IASB for the Standard or Interpretation.

And in the absence of any specific Standard or an Interpretation for any item, management must use its judgment in developing and applying an accounting policy that results in information that is relevant and reliable.

According to provisions of IAS 8 **while making that judgment**, management must refer to, and consider the applicability of, the following sources in **descending order:**

- ❖ the requirements and guidance in IASB standards and interpretations dealing with similar and related issues; and
- ❖ the definitions, recognition criteria, and measurement concepts for assets, liabilities, income, and expenses in the Framework.
- ❖ Management may also consider the most recent pronouncements of other standard-setting bodies that use a similar conceptual framework to develop accounting standards, other accounting literature, and accepted industry practices, to the extent that these do not conflict with the sources in paragraph 11

Consistency of accounting policies: IAS 8 prescribes that an entity shall select and apply its accounting policies consistently for similar transactions, other events, and conditions unless a Standard or an Interpretation specifically requires or permits categorization of items for which different policies may be appropriate. If a Standard or an Interpretation requires or permits such categorization, an appropriate accounting policy shall be selected and applied consistently to each category

Can an entity change its accounting policies:

An entity is permitted to change an accounting policy only if the change is required by a standard or interpretation; or

results in the financial statements providing reliable and more relevant information about the effects of transactions, other events, or conditions on the entity's financial position, financial performance, or cash flows.

If a change in accounting policy is required by a new IASB standard or interpretation, the change is accounted for as required by that new pronouncement or, if the new pronouncement does not include specific transition provisions, **then the change in accounting policy is applied retrospectively.**

Retrospective application means adjusting the opening balance of each affected component of equity for the earliest prior period presented and the other comparative amounts disclosed for each prior period presented as if the new accounting policy had always been applied.

However, if it is impracticable to determine either the period-specific effects or the cumulative effect of the change for one or more prior periods presented, the entity shall apply the new accounting policy to the carrying amounts of assets and liabilities as at the beginning of the earliest period for which retrospective application is practicable, which may be the current period, and shall make a corresponding adjustment to the opening balance of each affected component of equity for that period.

Also, if it is impracticable to determine the cumulative effect, at the beginning of the current period, of applying a new accounting policy to all prior periods, the entity shall adjust the comparative information to **apply the new accounting policy prospectively** from the earliest date practicable.

Changes in accounting estimates: IAS 8 prescribes that the effect of a change in an accounting estimate shall be recognized prospectively by including it in profit or loss in:

  ➢ the period of the change, if the change affects that period only, or
  ➢ the period of the change and future periods, if the change affects both.
  ➢ However, to the extent that a change in an accounting estimate gives rise to changes in assets and liabilities, or relates to an item of equity, it is recognized by adjusting the carrying amount of the related asset, liability, or equity item in the period of the change.

IAS 10- Events After the Reporting Period

IAS 10 Events After the Reporting Period contains guidelines for how events after the end of the reporting period should be treated in the financial statements.

What is an Event after the reporting period: An event, which could be favourable or unfavourable, that occurs between the end of the reporting period and the date that the financial statements are authorized for issue.

What is an Adjusting event: An event after the reporting period that provides further evidence of conditions that existed at the end of the reporting period, including an event that indicates that the going concern assumption in relation to the whole or part of the enterprise is not appropriate.

What is a non-adjusting event: An event after the reporting period that is indicative of a condition that arose after the end of the reporting period.

Accounting Treatment of Events after balance sheet date: Adjust financial statements for adjusting events - events after the balance sheet date that provide further evidence of conditions that existed at the end of the reporting period, including events that indicate that the going concern assumption in relation to the whole or part of the enterprise is not appropriate.

Do not adjust for non-adjusting events - events or conditions that arose after the end of the reporting period.

If an entity declares dividends after the reporting period, the entity shall not recognize those dividends as a liability at the end of the reporting period. That is a non-adjusting event.

What if Going concern issues arising after the end of the reporting period?

An entity shall not prepare its financial statements on a going concern basis if management determines after the end of the reporting period either that it intends to liquidate the entity or to cease trading, or that it has no realistic alternative but to do so.

IAS 34 Interim Financial Reporting:

A set of financial statements (Complete or condensed) for a period shorter than a financial year is termed as an Interim Financial report.

Which entity should publish an interim financial report is not a subject matter of IAS-34. It is generally governed by local laws and government regulations applicable for the entity i.e. IAS 34 applies if an entity using IFRS Standards in its annual financial statements and that entity publishes an interim financial report that asserts compliance with IFRS Standards.

IAS 34 prescribes the minimum content to be reported by any entity in its interim financial report. It also specifies the accounting recognition and measurement principles applicable to an interim financial report.

Presentation of Financial Reporting as per IAS-34:

Minimum content to be reported in Interim financial report: IAS 34 Prescribes if an entity publishes an interim financial report (which asserts compliance with IFRS Standards) then it should contain

> ➢ A set of condensed financial statements for the current period and comparative prior period information, ie statement of financial position,
> ➢ statement of comprehensive income,
> ➢ statement of cash flows,
> ➢ statement of changes in equity,
> ➢ and selected explanatory notes.

However, In some cases, a statement of financial position at the beginning of the prior period is also required. Generally, information available in the entity's most recent annual report is not repeated or updated in the interim report. The interim report deals with changes since the end of the last annual reporting period.

The interim financial report should be prepared by applying the same accounting policies as in the most recent annual report, or special disclosures are required if an accounting policy is changed.

Assets and liabilities are recognized and measured for interim reporting based on information available on a year-to-date basis.

Like in annual financial statements, measurements in interim financial reports are also based on reasonable estimates, indeed while preparing interim financial reports generally greater use of estimation methods are required than annual financial statements.

IFRS 3 Business Combinations

IFRS 3 Business Combinations prescribes the guidelines for accounting when an acquirer obtains control of a business (e.g., an acquisition or merger). Normally accounting for business combinations is done by applying the **'acquisition method'**, which generally requires assets acquired and liabilities assumed to be measured at their fair values at the acquisition date.

In short IFRS 3 establishes principles and requirements for how an acquirer in a business combination:

> - should recognize and measures in its financial statements the assets and liabilities acquired, and any interest in the acquiree held by other parties;
> - should recognize and measures the goodwill acquired or a gain from a bargain purchase; and
> - determines the disclosure to be made, to enable users of the financial statements, to evaluate the nature and financial effects of the business combination.

Key definitions under IFRS 3

Business Combination: A transaction or other event in which an acquirer obtains control of one or more businesses. Transactions sometimes referred to as 'true mergers' or 'mergers of equals' are also business combinations as that term is used in

Business: An integrated set of activities and assets that is capable of being conducted and managed for the purpose of providing goods or services to customers, generating investment income (such as dividends or interest) or generating other income from ordinary activities*

Acquisition Date: The date on which the acquirer obtains control of the acquiree

Acquirer: The entity that obtains control of the acquiree

Acquiree: The business or businesses that the acquirer obtains control of in a business combination

Core Principle of IFRS 3:

Step 1: Acquirer should measure the cost of the acquisition at the fair value of the consideration paid

Step 2: Allocates this cost to the acquired identifiable assets and liabilities based on their fair values

Step 3: Allocates the rest of the cost to goodwill, if any, or recognizes any excess of acquired assets and liabilities over the consideration paid (a 'bargain purchase') in profit or loss immediately.

Step 4: Discloses information that enables users to evaluate the nature and financial effects of the acquisition.

Method of accounting for business combinations: Accounting for all business combinations is done using the acquisition method

Following Steps are followed while applying the acquisition method
Identification of the 'acquirer'

➢ Determination of the 'acquisition date'
➢ Recognition and measurement of the identifiable assets acquired, the liabilities assumed and any non-controlling interest (NCI, formerly called minority interest) in the acquiree
➢ Recognition and measurement of goodwill or a gain from a bargain purchase

IFRS 8 Operating Segments

IFRS 8 requires an entity whose debt or equity securities are publicly traded to disclose information to enable users of its financial statements to evaluate the nature and financial effects of the different business activities in which it engages and the different economic environments in which it operates. It specifies how an entity should report information about its operating segments in annual financial statements and interim financial reports. It also sets out requirements for related disclosures about products and services, geographical areas, and major customers.

IFRS 8 applies to the separate or consolidated financial statements of an entity whose debt or equity instruments are traded in a public market or that files, or is in the process of filing, its (consolidated) financial statements with a securities commission or other regulatory organization for the purpose of issuing any class of instruments in a public market

However, when both separate and consolidated financial statements for the parent are presented in a single financial report, segment information need be presented only based on the consolidated financial statements

What is Operating segments

IFRS 8 defines an operating segment as a component of an entity that engages in business activities from which it may earn revenues and incur expenses and whose operating results are reviewed regularly by the entity's chief operating decision-maker to make decisions about resources to be allocated to the segment and assess its performance and for which discrete financial information is available

What are Reportable segments

IFRS 8 requires an entity to report financial and descriptive information about its reportable segments. Reportable segments are operating segments or aggregations of operating segments that meet specified criteria as given bellow

* ❖ its reported revenue, from both external customers and intersegment sales or transfers, is 10 percent or more of the combined revenue, internal and external, of all operating segments, or
* ❖ the absolute measure of its reported profit or loss is 10 percent or more of the greater, in absolute amount, of (i) the combined reported profit of all operating segments that did not report a loss and (ii) the combined reported loss of all operating segments that reported a loss, or
* ❖ its assets are 10 percent or more of the combined assets of all operating segments.
* ❖ Two or more operating segments may be aggregated into a single operating segment if aggregation is consistent with the core principles of the standard, the segments have similar economic characteristics and are similar in various prescribed respects.
* ❖ If the total external revenue reported by operating segments constitutes less than 75 percent of the entity's revenue, additional operating segments

must be identified as reportable segments (even if they do not meet the quantitative thresholds set out above) until at least 75 percent of the entity's revenue is included in reportable segments.

IFRS 10 Consolidated Financial Statements

Consolidated financial statements are financial statements that present the assets, liabilities, equity, income, expenses, and cash flows of a parent and its subsidiaries as those of a single economic entity.

IFRS 10 establishes principles for presenting and preparing consolidated financial statements when an entity controls one or more other entities. IFRS 10:

- requires an entity (the parent) that controls one or more other entities (subsidiaries) to present consolidated financial statements;
- defines the principle of control, and establishes control as the basis for consolidation;
- sets out how to apply the principle of control to identify whether an investor controls an investee and therefore must consolidate the investee;
- sets out the accounting requirements for the preparation of consolidated financial statements; and
- defines an investment entity and sets out an exception to consolidating particular subsidiaries of an investment entity.

IFRS 10 also prescribes the circumstances when Investment entities get exemption from the consolidation of financials

Key definitions of IFRS 10

Control of an investee: An investor controls an investee when the investor is exposed, or has rights, to variable returns from its involvement with the investee and has the ability to affect those returns through its power over the investee

Parent: A parent is an entity that **controls** one or more entities

Power: Existing rights that give the current ability to direct the relevant activities

Relevant Activity: Relevant Activities are Activities of the investee that significantly affect the investee's returns

What is Control:

An investor controls an investee if and only if the investor has all of the following elements

- ➤ power over the investee, i.e. the investor has existing rights that give it the ability to direct the relevant activities (the activities that significantly affect the investee's returns)
- ➤ exposure, or rights, to variable returns from its involvement with the investee
- ➤ the ability to use its power over the investee to affect the number of the investor's returns.

procedure to consolidate the financials

The broad procedure of consolidation of financials are as given below

- ❖ First combine like items of assets, liabilities, equity, income, expenses, and cash flows of the parent with those of its subsidiaries
- ❖ offset (eliminate) the carrying amount of the parent's investment in each subsidiary and the parent's portion of equity of each subsidiary
- ❖ eliminate intragroup transactions in full
- ❖ Unless impracticable, the parent and subsidiaries are required to have the same reporting dates, however, where reporting dates are different, the difference between the date of the subsidiary's financial statements and that of the consolidated financial statements shall be no more than three months
- ❖ Non-controlling interests are shown separately from the equity of the owners of the parent.
- ❖ The reporting entity also attributes total comprehensive income to the owners of the parent and the non-controlling interests even if this results in the non-controlling interests having a deficit balance.

IFRS 11 Joint Arrangements

IFRS 11 establishes principles for financial reporting by entities that have an interest in arrangements that are controlled jointly (joint arrangements).

What is the core principle of IFRS 11: The core principle is a party to a joint arrangement determines the type of joint arrangement in which it is involved by assessing its rights and obligations and accounts for those rights and obligations in accordance with that type of joint arrangement.

Key definitions of IFRS 11

Joint Arrangement: An arrangement in which two or more parties have joint control

Joint Control: The contractually agreed sharing of control of an arrangement, which exists only when decisions about the relevant activities require the unanimous consent of the parties sharing control

Joint Operation: A joint arrangement whereby the parties that have joint control of the arrangement have rights to the assets, and obligations for the liabilities, relating to the arrangement

Joint Venture: A joint arrangement whereby the parties that have joint control of the arrangement have rights to the net assets of the arrangement

Joint Venturer: A party to a joint venture that has joint control of that joint venture

Party to Joint Arrangement: An entity that participates in a joint arrangement, regardless of whether that entity has joint control of the arrangement

Separate Vehicles: A separately identifiable financial structure, including separate legal entities or entities recognized by statute, regardless of whether those entities have a legal personality

Accounting for Joint arrangement in Separate Financial Statements

The accounting for joint arrangements in an entity's separate financial statements depends on the involvement of the entity in that joint arrangement and the type of the joint arrangement:

How D365 ERP supports IAS 1 & others

Refer to Chapter 11 & 12 of this book to understand more on how this ERP supports IAS-1 and other presentation or disclosure related requirements as per IAS 7, IAS8, IAS 10, IAS 34, IFRS 3, IFRS 8, IFRS 10, IFRS 11 with some inbuilt capabilities and with a report designer tool. Here is the summary:

❖ D365 ERP comes with some out-of-the-box financial reports which can be modified further as per the requirement of the company and the country. As we said before, each country can have its guidelines on the format.

❖ 'Financial reporting' and 'Power BI' are the native tools of Microsoft D365 which are being widely used to design financial statements nowadays.

❖ Some companies maintain their reporting tool (EPM tools) and they just need data from D365 which is also possible in several ways.

❖ D365 supports maintaining financial calendars, defining reporting periods, closing books as per the company's requirements.

🔅 We want to reiterate here that IAS-1 prescribes the purpose and structure of financial statement, but it *does not prescribe any standard format* for presentation of financial statements.

Each company is free to choose what format they want to adopt as per their *local legislations and guidelines,* and they will still be compliant with IFRS 1

IAS 2 — Inventories

Overview of IAS 2

The objective of IAS 2 is to prescribe the accounting treatment for inventories. It provides guidance for determining the cost of inventories and for subsequently recognizing an expense, including any write-down to net realizable value. It also guides the cost formulas that are used to assign costs to inventories.

Scope of IAS 2

Inventories **include** assets held for sale in the ordinary course of business (**finished goods**), assets in the production process for sale in the ordinary course of business (**work in process**), and materials and supplies that are consumed in production (**raw materials**).

This Standard **does not apply to the following items** even if having characteristics of Inventory

- Work in process arising under construction contracts – IAS11 Construction
- Financial Instruments – IAS 39 Financial Instruments: Recognition and measurement
- Biological assets related to agricultural activity and agricultural produce at the point of harvest - IAS 41 Agriculture.

Also, while the following are within the scope of the standard, IAS 2 **does not apply** to the measurement of inventories held by: [IAS 2.3]

- *__producers of agricultural and forest products, agricultural produce__* after harvest, and minerals and mineral products, to the extent that they are measured at net realizable value (above or below cost) in accordance with well-established practices in those industries. When such inventories are measured at net realizable value, changes in that value are recognized in profit or loss in the period of the change

- *__commodity brokers and dealers__ who measure their inventories at fair value less costs to sell*. When such inventories are measured at fair value less costs to sell, changes in fair value less costs to sell are recognized in profit or loss in the period of the change.

The fundamental principle of IAS 2
Inventories are required to be stated at the lower of cost and net realizable value.

Measurement of inventories
Cost **should include** all:

- costs of purchase (including taxes, transport, and handling) net of trade discounts received
- costs of conversion (including fixed and variable manufacturing overheads) and
- other costs incurred in bringing the inventories to their present location and condition
- Normally Borrowing Cost is not included in the cost of inventory however para 4 of **IAS 23** Borrowing Costs identifies some limited circumstances where borrowing costs (interest) can be included in the cost of inventories that meet the definition of a qualifying asset

Inventory cost **should not include** the following expenses
> ➢ abnormal waste
> ➢ storage costs
> ➢ administrative overheads unrelated to production
> ➢ selling costs
> ➢ foreign exchange differences arising directly on the recent acquisition of inventories invoiced in a foreign currency
> ➢ interest cost when inventories are purchased with deferred settlement terms.

o IAS-2 permits standard cost and retail methods for the measurement of cost **provided that** the results approximate to the actual cost.
o For inventory items that are not interchangeable, specific costs are attributed to the specific individual items of inventory
o For interchangeable items, IAS 2 allows the **FIFO** or **weighted average cost** formulas, however, _LIFO is not allowed_

Note- Inventories having similar characteristics should be measured using the same cost formula. For groups of inventories that have different characteristics, different cost formulas may be justified.

Expense recognition-
Expense should be recognized in the Profit & Loss Account in the following circumstances

    a.   When Inventories are sold, and revenue is recognized in P/L

    b.   There is a write-down of inventories to NRV or Inventory loss

Sale of Inventory:

When inventories are sold, the **carrying amount** of those inventories shall be recognized as an expense in the period in which the related revenue is recognized. (It is known as Cost of Goods Sold- COGS)

Write-down to net realizable value (NRV) / Inventory Loss

NRV is the estimated selling price in the ordinary course of business, less the estimated cost of completion and the estimated costs necessary to make the sale.

- The amount of any write-down of inventories to net realizable value and all losses of inventories shall be recognized as an expense in the period the write-down or loss occurs.

- The amount of any reversal of any write-down of inventories, arising from an increase in net realizable value, shall be recognized as a reduction in the number of inventories recognized as an expense in the period in which the reversal occurs

How D365 ERP handles IAS 2

Refer to Chapter 10A how D365 ERP supports IAS 2 in detail and here is the summary:

Measurement of Inventory

- Landed Cost Module of D365 helps to calculate the landed cost/accurate carrying cost (including all freight, insurance, etc.)
- BOM in the production module helps in the automatic accumulation of cost of conversion on the finished product
- Cost Sheet in Production Module helps in valuation of produced items, absorption of indirect cost.

Valuation Method

D365 supports all valuation methods which are permissible by IFRS e.g., Weighted Average, FIFO, and standard costing.

Expense Recognition

Cost of Goods Sold- D365 ERP has an automatic recognition of expense when a product is sold. It is known as Cost of Goods Sold (COGS). COGS is calculated based on the 'Inventory Valuation Method' selected by the organization for that product. (See 'Chapter- 9 Order to Cash in D365 ERP')

Net Realizable Value (NRV) – NRV is calculated outside of the system, and it is up to the management of the company. D365 can help to adjust the value of the inventory if the NRV is less than the carrying cost of the inventory and the management decides to write down the value.

IAS 21 — The Effects of Changes in Foreign Exchange Rates

Overview of IAS 21

The objective of IAS 21 is to prescribe how to include foreign **currency transactions** and **foreign operations** in the financial statements of an entity and how to *translate financial statements into a presentation currency.*

The fundamental principle of IAS 21

The principal issues are which exchange rate(s) to use and how to report the effects of changes in exchange rates in the financial statements.

Key definitions from IAS 21

Functional currency: the currency of the primary economic environment in which the entity operates.

Monetary items: The essential feature of a monetary item is a right to receive (or an obligation to deliver) a fixed or determinable number of units of currency. Examples include pensions and other employee benefits to be paid in cash; provisions that are to be settled in cash; and cash dividends that are recognized as a liability. Similarly, a contract to receive (or deliver) a variable number of the entity's own equity instruments or a variable amount of assets in which the fair value to be received (or delivered) equals a fixed or determinable number of units of currency is a monetary item.

Non-Monetary items: Conversely, the essential feature of a non-monetary item is the absence of a right to receive (or an obligation to deliver) a fixed or determinable number of units of currency. Examples include amounts prepaid for goods and services (eg prepaid rent); goodwill; intangible assets; inventories; property, plant and equipment; and provisions that are to be settled by the delivery of a nonmonetary asset.

Presentation currency: the currency in which financial statements are presented.

Exchange difference: the difference resulting from translating a given number of units of one currency into another currency at different exchange rates.

Foreign operation: a subsidiary, associate, joint venture, or branch whose activities are based in a country or currency other than that of the reporting entity.

(A) *Translating Financial Statement into Presentation Currency*

The results and financial position of an entity whose functional currency is not the currency of a hyperinflationary economy are translated into a different presentation currency using the following procedures: [IAS 21.39]

- assets and liabilities for each balance sheet presented (including comparatives) are translated at the closing rate at the date of that balance sheet. This would include any goodwill arising on the acquisition of a foreign operation and any fair value adjustments to the carrying amounts of assets and liabilities arising on the acquisition of that foreign operation are treated as part of the assets and liabilities of the foreign operation [IAS 21.47];

- income and expenses for each income statement (including comparatives) are translated at exchange rates at the dates of the transactions, and
- all resulting exchange differences are recognized in other comprehensive income.

(B) *Foreign Currency Transactions*

A foreign currency transaction should be recorded initially at the rate of exchange at the date of the transaction (use of averages is permitted if they are a reasonable approximation of actual).

At each subsequent balance sheet date:

- Foreign currency monetary items should be reported using the closing rate
- non-monetary items carried at historical cost should be reported using the exchange rate at the date of the transaction
- non-monetary items carried at fair value should be reported at the rate that existed when the fair values were determined

Exchange differences arising when monetary items are settled or when monetary items are translated at rates different from those at which they were translated when initially recognized or in previous financial statements are reported in profit or loss in the period, with one exception.

The exception is that exchange differences arising on monetary items that form part of the reporting entity's net investment in a foreign operation are recognized, in the consolidated financial statements that include the foreign operation, in other comprehensive income; they will be recognized in profit or loss on disposal of the net investment.

Basic steps for translating foreign currency amounts into the functional currency

Steps apply to a stand-alone entity, an entity with foreign operations (such as a parent with foreign subsidiaries), or a foreign operation (such as a foreign subsidiary or branch).

- the reporting entity determines its functional currency
- the entity translates all foreign currency items into its functional currency
- the entity reports the effects of such translation in accordance with this standard

As regards a monetary item that forms part of an entity's investment in a foreign operation, the accounting treatment in consolidated financial statements should not be dependent on the currency of the monetary item. [IAS 21.33] Also, the accounting should not depend on which entity within the group conducts a transaction with the foreign operation. [IAS 21.15A] If a gain or loss on a non-monetary item is recognized in other comprehensive income (for example, a property revaluation under IAS 16), any foreign exchange component of that gain or loss is also recognized in other comprehensive income. [IAS 21.30]

Disposal of a foreign operation

When a foreign operation is disposed of, the cumulative amount of the exchange differences recognized in other comprehensive income and accumulated in the separate component of equity relating to that foreign operation shall be recognized in profit or loss when the gain or loss on disposal is recognized.

Tax effects of exchange differences

These must be accounted for using **IAS 12 Income Taxes**.

Hyperinflationary Economy

Special rules apply for translating the results and financial position of an entity whose functional currency is the currency of a **hyperinflationary economy** into a different presentation currency.

Where the foreign entity reports in the currency of a hyperinflationary economy, the financial statements of the foreign entity should be restated as required by IAS 29 Financial Reporting in Hyperinflationary Economies, before translation into the reporting currency.

In 2020, all those entities whose functional currency was the currency of one of the following countries were considered hyperinflationary economies-

- Argentina
- Islamic Republic of Iran
- Lebanon
- South Sudan
- Sudan
- Venezuela
- Zimbabwe

This list might change every year depending upon the economic outlook and inflation situation of the countries.

How D365 ERP handles IAS 21

D365 ERP has strong capabilities to manage the expectations of IAS 21.

(A) *Translating Financial Statement into Presentation Currency*

Investments in foreign operations can be recognized in Dynamics 365 for Operations as separate financial statements of the reporting entity.

We always recommend using the 'Financial Reporter' tool of D365 to translate the foreign currency entities into the presentation currency. *'Currency Translation Type' can be defined on a chart of accounts of the foreign entity for each ledger account (see the screen below). All the options are available which are prescribed in IAS 21.*

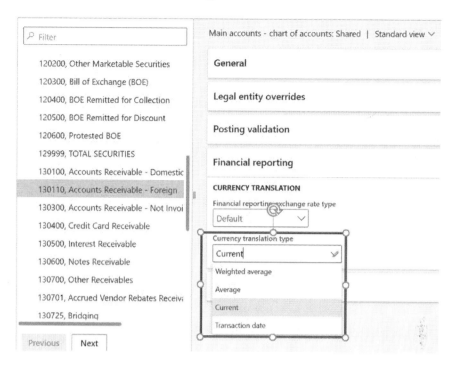

For consolidation companies, the transaction rate, average exchange rate, or closing rate can be applied to accounts. The difference that is calculated to balance when you translate to the reporting currency will be identified as the currency translation adjustment amount and will be presented appropriately in the financial statements.

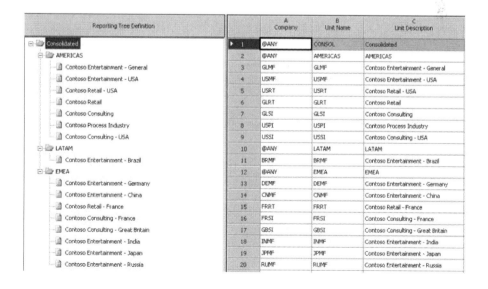

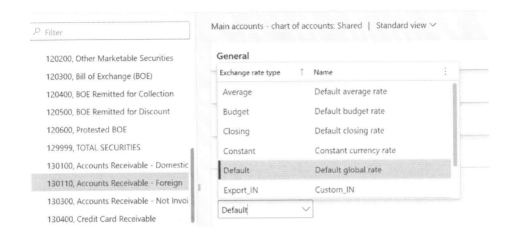

The best feature of 'D365 Financial reports' is that you can view the reports in any currency (NOT only in the reporting currency or functional currency) at the run time and the rate will be applied automatically from the configuration we discussed above.

CONTOSO ENTERTAINMENT SYSTEM USA

(B) *Foreign Currency Transactions*

Refer to <u>Chapter 6, heading 6.3</u>, how D365 ERP supports IAS 21 in detail, and here is the summary:

Monetary Items-

Generally, <u>Accounts receivables, Accounts Payable, Bank balances</u> are considered monetary items in a business. D365 has a feature to run the revaluation at the month-end

Non-monetary Items

Fixed assets that are bought in foreign currency are an example of non-monetary items and those can be retained at the historical exchange rate i.e. there is no need to run the revaluation process. See the following screen of D365 where we can define at the chart of accounts level which exchange rate will be used for the revaluation and if the account will be considered in the currency revaluation process or not.

IAS 16 — Property, Plant and Equipment

Overview of IAS 16

IAS 16 Property, Plant, and Equipment outline the accounting treatment for most types of property, plant, and equipment. Property, plant, and equipment are initially measured at their cost, subsequently measured either using a cost or revaluation model and depreciated so that their depreciable amount is allocated on a systematic basis over its useful life.

Scope

IAS 16 applies to the accounting for property, plant, and equipment, except where another standard requires or permits differing accounting treatments, for example:

A. assets classified as held for sale in accordance with IFRS 5 *Non-current Assets Held for Sale and Discontinued Operations*

B. biological assets related to agricultural activity accounted for under IAS 41 *Agriculture*
C. exploration and evaluation assets recognized in accordance with IFRS 6 *Exploration for and Evaluation of Mineral Resources*
D. mineral rights and mineral reserves such as oil, natural gas, and similar non-regenerative resources.

Recognition

Items of property, plant and equipment should be recognized as assets when it is probable that **future economic benefits** associated with the asset will flow to the entity, and the **cost of the asset can be measured reliably**.

These costs include costs incurred initially to acquire or construct an item of property, plant, and equipment and costs incurred subsequently to add to, replace part of, or service it.

Initial measurement

An item of property, plant and equipment should initially be recorded at cost. Cost includes all costs necessary to bring the asset to working condition for its intended use. This would include not only its original purchase price but also costs of site preparation, delivery and handling, installation, related professional fees for architects and engineers, **and the estimated cost of dismantling and removing the asset and restoring the site**

Measurement Subsequent to initial recognition

IAS 16 permits two accounting models:
  ➢ **Cost model.** The asset is carried at cost less accumulated depreciation and impairment.
  ➢ **Revaluation model.** The asset is carried at a revalued amount, being its fair value at the date of revaluation less subsequent depreciation and impairment, provided that fair value can be measured reliably. Under the revaluation model, evaluations should be carried out on each balance sheet date. If an item is revalued, the entire class of assets to which that asset belongs should be revalued.
  ➢ Revalued assets are depreciated in the same way as under the cost model If a revaluation increases in value, it should be credited to other comprehensive income and accumulated in equity under the heading "revaluation surplus" unless it represents the reversal of a revaluation decrease of the same asset previously recognized as an expense, in which case it should be recognized in profit or loss.

> ➤ A decrease arising as a result of a revaluation should be recognized as an expense to the extent that it exceeds any amount previously credited to the revaluation surplus relating to the same asset.
> ➤ When a revalued asset is disposed of, any revaluation surplus may be transferred directly to retained earnings, or it may be left in equity under the heading revaluation surplus. The transfer to retained earnings should not be made through profit or loss.

Depreciation (cost and revaluation models)

For all depreciable assets:

The depreciable amount (cost less residual value) should be allocated on a systematic basis over the asset's useful life

The residual value and the useful life of an asset should be reviewed at least at each financial year-end and, if expectations differ from previous estimates, any change is accounted for **prospectively as a change in estimate under IAS 8.**

The depreciation method used should reflect the pattern in which the asset's economic benefits are consumed by the entity

The depreciation method should be reviewed at least annually and, if the pattern of consumption of benefits has changed, the depreciation method should be changed **prospectively as a change in estimate under IAS 8.**

Depreciation should be charged to profit or loss unless it is included in the carrying amount of another asset. Depreciation begins when the asset is available for use and continues until the asset is derecognized, even if it is idle.

Recoverability of the carrying amount

IAS 16 *Property, Plant, and Equipment* require impairment testing and, if necessary, recognition for property, plant, and equipment. An item of property, plant, or equipment shall not be carried at more than a recoverable amount. The recoverable amount is the higher of an asset's fair value less costs to sell and its value in use.

Derecognition (retirements and disposals)

An asset should be removed from the statement of financial position on disposal or when it is withdrawn from use and no future economic benefits are expected from its disposal. The gain or loss on disposal is the difference between the proceeds and the carrying amount and should be recognized in profit and loss.

How D365 ERP handles IAS 16

D365 ERP has very strong capabilities to handle the requirements of IAS 16. Refer to **Chapter-8** of this book for more information where all the transactions are discussed in detail e.g., acquisition of the assets, acquisition adjustments, depreciation, revaluation, disposal of the assets.

We would like to reiterate here that there is an element of management decision when it comes to revaluation (similar to IAS36) of assets, any system will not be able to help in that for obvious reasons, otherwise, D365 ERP can handle all accounting treatment required by IAS 16.

IAS 36 — Impairment of Assets

Scope

To ensure that assets are carried at no more than their recoverable amount, and to define how recoverable amount is determined.

Applicability: IAS 36 applies to all assets except:

- inventories (IAS 2)
- assets arising from construction contracts (IAS 11)
- deferred tax assets (IAS 12)
- assets arising from employee benefits (IAS 19)
- financial assets (see IAS 39)
- investment property carried at fair value (IAS 40)
- agricultural assets carried at fair value (IAS 41)
- insurance contract assets (IFRS 4)
- non-current assets held for sale (IFRS 5)

Key definitions from IAS 36

Impairment loss: the amount by which the carrying amount of an asset or cash-generating unit exceeds its recoverable amount

Carrying amount: the amount at which an asset is recognized in the balance sheet after deducting accumulated depreciation and accumulated impairment losses

Recoverable amount: the higher of an asset's fair value less costs of disposal and its value in use

Fair value: the price that would be received to sell an asset or paid to transfer a liability in an orderly transaction between market participants at the measurement date (IFRS 13 Fair Value Measurement)

Value in use: the present value of the future cash flows expected to be derived from an asset or cash-generating unit

Identifying an asset that may be impaired:

At the end of each reporting period, an entity is required to assess whether there is any indication that an asset may be impaired (i.e., carrying amount may be higher than its recoverable amount). IAS 36 has a list of external and internal indicators of impairment. If there is an indication that an asset may be impaired, then the asset's recoverable amount must be calculated.

The fundamental principle of IAS 36

The fundamental principle is that an asset should not be carried on the balance sheet at an amount that is above its recoverable amount

The recoverable amounts of the following types of intangible assets are measured annually whether or not there is any indication that it may be impaired.

➤ an intangible asset with an indefinite useful life
➤ an intangible asset not yet available for use
➤ goodwill acquired in a business combination

Indications of impairment can be external like.

> - market value declines,
> - negative changes in technology, markets, economy, or laws
> - increases in market interest rates
> - net assets of the company higher than market capitalization

or the indications can be internal like.

> - obsolescence or physical damage
> - the asset is idle, part of a restructuring, or held for disposal
> - worse economic performance than expected
> - for investments in subsidiaries, joint ventures, or associates, the carrying amount is higher than the carrying amount of the investee's assets, or a dividend exceeds the total comprehensive income of the investee

These lists are inclusive, there may be many other indicators of impairment

Cash-generating units

> - The recoverable amount should be determined for the individual asset, if possible.
> - If it is not possible to determine the recoverable amount for the individual asset, then determine the recoverable amount for the asset's cash-generating unit **(CGU)**. [The CGU is the smallest identifiable group of assets that generates cash inflows that are largely independent of the cash inflows from other assets or groups of assets.]
> - Impairment of goodwill: Goodwill should be tested for impairment annually.
>
> - To test for impairment, goodwill must be allocated to each of the acquirer's cash-generating units, or groups of cash-generating units, that are expected to benefit from the synergies of the combination, irrespective of whether other assets or liabilities of the acquiree are assigned to those units or groups of units.
>
> - Each unit or group of units to which the goodwill is so allocated shall represent the lowest level within the entity at which the goodwill is monitored for internal management purposes; and not be larger than an

operating segment determined in accordance with IFRS 8 Operating Segments.

A cash-generating unit to which goodwill has been allocated shall be tested for impairment at least annually by comparing the carrying amount of the unit, including the goodwill, with the recoverable amount of the unit

> If the recoverable amount of the unit exceeds the carrying amount of the unit, the unit and the goodwill allocated to that unit is not impaired
> If the carrying amount of the unit exceeds the recoverable amount of the unit, the entity must recognize an impairment loss.

The impairment loss is allocated to reduce the carrying amount of the assets of the unit (group of units) in the following order

> first, reduce the carrying amount of any goodwill allocated to the cash-generating unit (group of units); and
> then, reduce the carrying amounts of the other assets of the unit (group of units) pro rata on the basis.

Accounting Entry for Impairment Loss

Debit - Loss on Impairment

Credit - Accumulated Impairment Loss

(if there is no separate ledger for Accumulated Impairment loss then it can be credited to Accumulated Depreciation Account)

Reversal of an impairment loss

> Same approach as for the identification of impaired assets: assess at each balance sheet date whether there is an indication that an impairment loss may have decreased. If so, calculate the recoverable amount.

> The increased carrying amount due to reversal should not be more than what the depreciated historical cost would have been if the impairment had not been recognized.
> Reversal of an impairment loss is recognized in the profit or loss unless it relates to a revalued asset
> Reversal of an impairment loss for goodwill is prohibited.

How IAS 36 is practically used

Practically, to calculate the carrying amount of each asset covered in IAS-36, a separate assets memorandum account chart is prepared where accumulated depreciation/ accumulated Impairment loss is apportioned to individual assets. For balance sheet presentation purposes group-wise (Plant & Machinery, Furniture, cars, etc.) consolidated value (cost or fair value) and accumulated depreciation of tangible/ intangible Assets are shown.))

How D365 ERP handles IAS 36

D365 supports the booking of an impairment loss or gain through the revaluation process in the Fixed Assets Module. There is a decision element in impairment that is performed outside of any system by the management or the qualified valuers.

Some new features have been introduced for the Japan Localization package of D365 ERP where many of the IAS 36 related requirements can be handled out of the box.

  ➢ Assigning the assets to CGU (Cash Generating Units)
  ➢ Impairment recognition test
  ➢ Measurement of the impairment amount
  ➢ Posting the impairment

Check the following link from Microsoft's official website.
Fixed asset impairment accounting on cash-generating units for Japan - Finance | Dynamics 365 | Microsoft Docs

IAS 40 — Investment Property

Scope

IAS 40 Investment Property applies to the accounting for property (land and/or buildings) held to earn rentals or for capital appreciation (or both).

Definition of investment property

Investment property is property (land or a building or part of a building or both) held (by the owner or by the lessee under a finance lease) to earn rentals or for capital appreciation or both

Examples of investment property:

> ➢ land held for long-term capital appreciation or held for an undetermined future use
> ➢ building leased out under an operating lease or held to be leased out under the operating lease
> ➢ Under construction property with an intention to hold it in future investment property

Exclusions: The following are not investment property and, therefore, are outside the scope of IAS 40

> ➢ Property covered in the definition of PPE
> ➢ property held for sale in the ordinary course of business or the process of construction of development for such sale i.e. covered under the definition of Inventory
> ➢ property being constructed or developed on behalf of third parties- Covered under IAS-11-Construction Contracts
> ➢ property leased out on finance lease

Recognition

When to Recognize: Investment property should be recognized as an asset when it is probable that the future economic benefits that are associated with the property will flow to the entity, and the cost of the property can be reliably measured.

Amount to Recognize: Initial measurement: Investment property is initially measured at cost, including transaction costs. Cost should not include start-up costs, abnormal waste, or initial operating losses incurred before the investment property achieves the planned level of occupancy.

Subsequent recognition: IAS 40 permits entities to choose between a fair value model, and a cost model.

One method must be adopted for all of an entity's investment property. An entity can change the method only if this results in a more appropriate presentation.

> ➤ Fair Value Model: Like in PPE Gains or losses arising from changes in the fair value of investment property must be included in net profit or loss for the period in which it arises.
> ➤ Cost model: After initial recognition, investment property is accounted for in accordance with the cost model as set out in IAS 16 Property, Plant and Equipment – cost less accumulated depreciation and less accumulated impairment losses.

Transfers to or from investment property classification: Transfers to, or from, investment property should only be made when there is a change in use, evidenced by facts.

Derecognition/ Disposal of Investment Property

An investment property should be derecognized on disposal or when the investment property is permanently withdrawn from use and no future economic benefits are expected from its disposal. The gain or loss on disposal should be calculated as the difference between the net disposal proceeds and the carrying amount of the asset and should be recognized as income or expense in the income statement/ profit & loss account.

How D365 ERP handles IAS 36

D365 ERP can record investment assets in the Fixed Assets module which is discussed in Chapter-8 of this book. The only difference between investment assets to the other fixed assets (IAS 16) will be that investment assets are not subject to depreciation. Also, these assets are disclosed differently as per IAS 1.

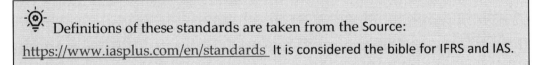

Definitions of these standards are taken from the Source:
https://www.iasplus.com/en/standards It is considered the bible for IFRS and IAS.

Visit our website www.satyakejriwal.com for more blogs and videos on IFRS.

Section-II
Microsoft D365 ERP
(Finance & Operations)

Finance and Operations

Dynamics 365

Note- The product is branded as Microsoft D365 ERP currently and it was also known as Dynamics AX 2012, AX2012 or Axapta in its previous versions. So, we have used some old version names as well wherever it was required.

6

General Ledger in D365 ERP

General Ledger is the most basic module for accounting in D365. It is quite simple to understand if you know the basics of accounting. There is not much setup required for accounts except chart of accounts, financial dimension, account structure, journal names, calendar, etc. I am underestimating the general ledger module by saying 'there is not much set up' and one reason is that I am from a finance background, and I have been doing it for many years now.

This module needs accounting knowledge, and it demands farsightedness from the consultants since here you finalize the chart of accounts and financial dimensions which is the backbone of financial reporting. The time invested in this discussion goes a long way when it comes to setting up the framework of reporting for the whole organization.

We will not talk about the chart of accounts and financial dimension again as we have already touched on it briefly in the previous chapters.

Sticking to the theme of this book, let's talk about the accounting entries which are possible under the General Ledger module and what is the rationale for those debit and credits.

6.1 General Journal

The general journal form is the first screen in D365 to capture a journal voucher. It is a very flexible voucher entry screen that you can expect in any accounting system.

Path: General Ledger → Journal Entries→ General Journal

First, you create the journal (aka batch number) and then you can create multiple lines and multiple vouchers in the same batch.

Following is the voucher entry screen in the journal. Again, it is a simple screen to capture the debit and credit of a transaction.

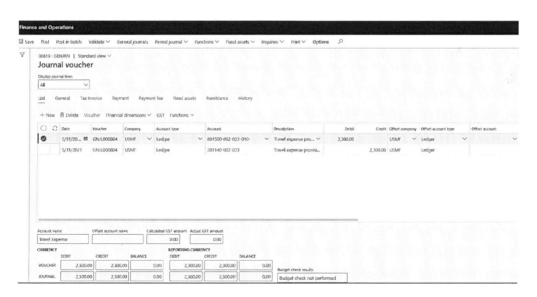

Following is the voucher after you post the journal.

- A voucher can be entered in a single line using an offset account or it can be multiple lines.

- Not all the ledger codes can be used directly in the journal. Some ledgers will be locked at the chart of accounts level for manual entry e.g. accounts receivable, accounts payable.

- Accounts and dimensions are entered in the same field generally delimited with '-'and it is another unique way to capture the dimension values.

- Debit and credit should be equal in the journal lines, otherwise, the voucher will not post. (This control is in line with the basic accounting principle of 'Dual Aspect Concept' which we read in chapter 1.)

What is **offset account** in the journal screen? That's the question asked by all of the new customers of D365, and I always give this comfort to them that it is a bit confusing, and they are not the first one to ask this.

So, in D365 you can enter a simple voucher in two ways, either you create two lines (As in the screen above), enter debit and credit in separate lines OR you just create one line for debit and the offset account will represent the credit side of that line.

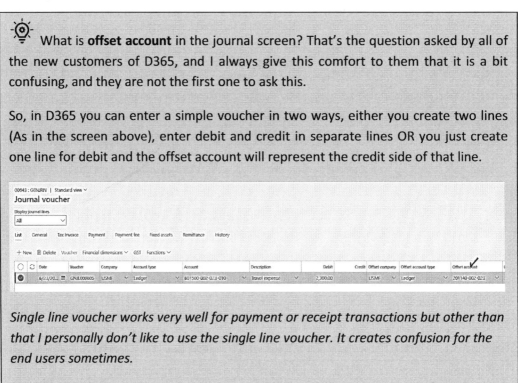

Single line voucher works very well for payment or receipt transactions but other than that I personally don't like to use the single line voucher. It creates confusion for the end users sometimes.

- Account type can be customer, vendor, project, fixed assets, bank, or Ledger. Technically, it means you can use the journal voucher screen for any transaction in the system. But will you ever do that? We won't do that on the general journal screen at least. There are other types of journals in D365 where we use other types of accounts. See the table below:

There are multiple variations of general journal which serve different purposes in D365. Here are the key journals which are generally used across all the industries:

Journal Name	Module	Purpose
Global General Journal	General Journal	Same as General Journal but ability to access all legal entities from the same screen
Allocation Journal	General Journal	It is used to allocate the expense based on the pre-defined allocation rules
Periodic Journal	General Journal	This journal is used as a template to reuse for recurring transactions.
Vendor Invoice Journal	Accounts Payable	To capture the non-inventory and non-purchase order invoices from the suppliers.
Global Invoice Journal	Accounts Payable	Same as Vendor Invoice Journal but ability to access all legal entities from the same screen
Vendor Payment Journal	Accounts Payable	To Capture the payment to the suppliers/Vendors
Customer Payment Journal	Accounts Receivable	To capture the payment from the customers.

Fixed Assets Journal	Fixed Assets	To capture the fixed assets capitalization, depreciation, write-off, revaluation, etc. transactions.

6.2 Inter-Company Accounting

When you have two legal entities in the same D365 environment, then there is a feature to automate the intercompany accounting. Generally, it applies to the scenario, such as daily journal vouchers, vendor invoice journals, and centralized payments.

> 💡 Please note that there is no limit on the number of companies or entities in D365 ERP which means that if a customer buys a license for D365, they can maintain any number of entities. Some of the accounting systems license on the basis of number of legal entities.

Path: General Ledger→ Posting Setup→Intercompany Accounting

In this setup, you are supposed to define a unique debit and credit account for each destination company e.g. I am incurring an expense from USMF for my sister concern DEMF. In this case, USMF is my originating company (where I am initiating the transaction from) and DEMF is my destination company.

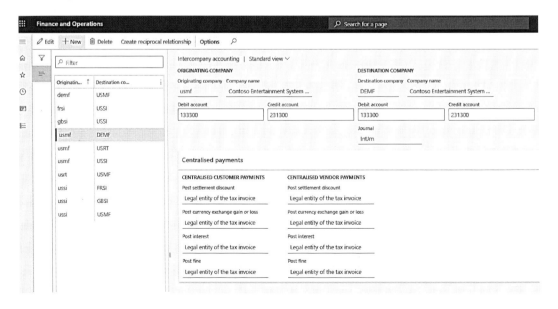

A typical inter-company transaction is captured in the same way as any other voucher except the company ID will be the destination company ID which means that we are hitting the ledger account of other legal entities directly without switching to that entity in the system. See the example below- 601500 is the expense account from DEMF legal entity but I have booked this invoice in USMF entity.

Now, there will be two separate but related accounting entries. The first voucher will be in USMF legal entity where the accounts payable will be credited (liability is booked in USMF) but the debit side (133300) is automatically picked up from the 'Inter-company' accounting set up as we discussed above.

The second voucher is in DEMF legal entity where 601500 account is debit and credit is again an automatic account (231300) from the 'inter-company accounting setup.

Here is our simple analysis of the accounting behind inter-company transactions-

Accounting Trigger in D365	Legal Entity	Type of Account	Ledger Account	Account	Debit	Credit	Accounting Rule	Logic for Debit/Credit	
Manual	USMF	Vendor	200110	Accounts Payable-Others		4000	Golden-1	Credit the increase in the liability	Set-1
Automatic	USMF	Ledger	133500	Intercompany Receivable DEMF/USMF	4000		Golden-1	Debit the increase in the asset	
Manual	DEMF	Ledger	601500	Travel Expense	4000		Golden-2	Debit the increase in the expense	Set-2
Automatic	DEMF	Ledger	231300	Intercompany Payable-USMF/DEMF		4000	Golden-1	Credit the increase in the liability	

> 💡 Highly recommended to use a unique main account for each legal entity. It will be a nightmare to reconcile the intercompany balances if all 'due to' and 'due from' balances are sitting in the same account. It will take a few new main accounts in chart of accounts, but it is worth creating it. ⏀
>
> I have worked for a client who had more than 15 related legal entities globally having their own local currencies. Due to lack of experience, we set up the same account for all legal entities and unfortunately, there was some balance mismatch after a few months which was almost impossible to catch due to currency translation and some ongoing bug in AX2012 earlier version.

6.3 Foreign Currency Revaluation

Foreign Currency Revaluation- Ledger
D365 support foreign currency revaluation at the following levels-

- General Ledger
- Customer Balance
- Vendor Balance
- Bank Balance

We will discuss primarily the ledger balance revaluation here.

<u>What accounts will be revalued?</u>
There is a flag on main accounts which is called 'Foreign Currency Revaluation'. Enabled this flag for those main accounts which need to be revalued.

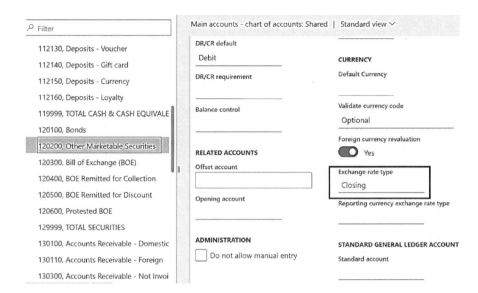

 A few points to remember-

- Only those accounts need to be flagged which will carry transaction in foreign currency.

- Don't flag the control accounts for customers, vendors and bank as there is a separate process to revalue then in AR, AP and bank module respectively.

What exchange rate type will be used for revaluation?

The exchange rate type is also defined on the ledger level (as you can see in the above screen), so you can define specific types of rates for each ledger. If nothing is set up on the ledger level, the rate will be taken from the rate type defined on the ledger set up in the General ledger set up. See the field 'Accounting currency exchange rate type' in the below screen.

 IAS 21 — The Effects of Changes in Foreign Exchange Rates

International Accounting Standard (IAS-21) require general ledger account balances in foreign currencies to be revalued using different exchange rate types (current, historical, average, etc.).

For example, one accounting convention requires following conversion rates-

- Assets and liabilities→ Current exchange rate,

- Fixed assets→Historical exchange rate, and

- Profit and loss accounts→ Monthly average.

Where to define the exchange rate loss/gain accounts?

The ledger accounts for foreign currency revaluation are defined on the following path:

Path: General Ledger→ Ledger Set up → Ledger

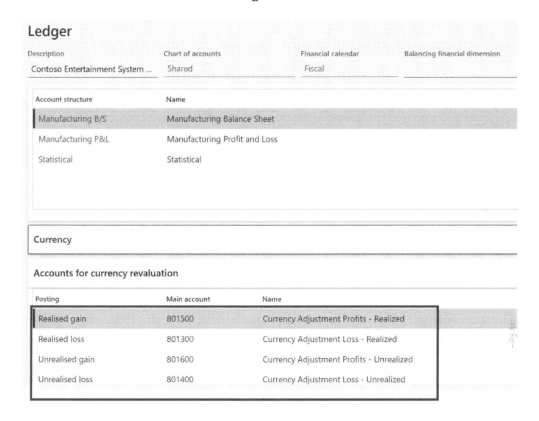

If the company decides to maintain currency wise ledger accounts, there is also an option to define it per currency on the following path:

Path: General Ledger→ Posting Setup → Currency Revaluation Accounts

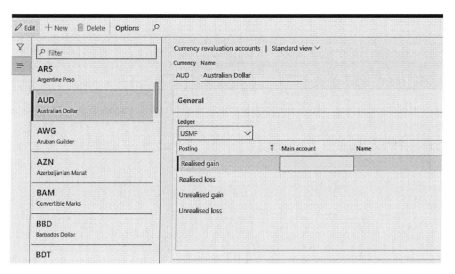

How to run the revaluation process and accounting impact?
Path: General Ledger→ Periodic → Foreign Currency Revaluation

This is the process where you can give the currency, legal entity, date range, date of rate, and select which accounts you want to revalue.

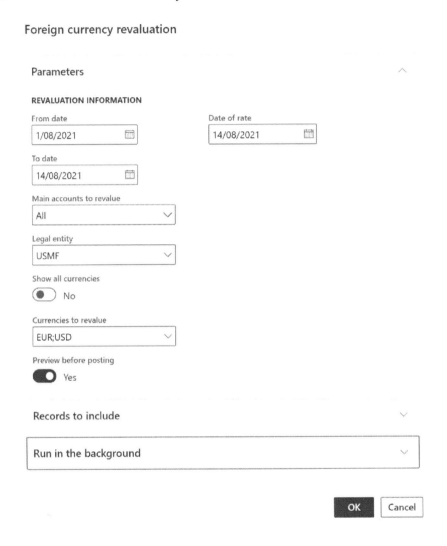

Here is the sample voucher created from the revaluation process. You will notice that only two main accounts are covered in this process for obvious reasons.

Voucher transactions

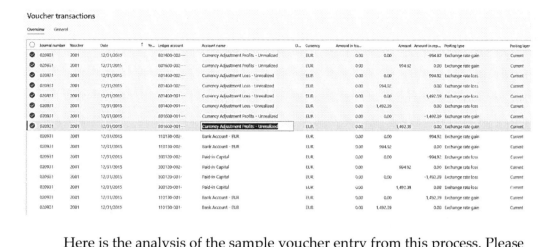

Here is the analysis of the sample voucher entry from this process. Please note when you run the GL revaluation, it will always hit the 'unrealized loss or gain' account. 'Realized loss/gain' account is used only for AR, AP, and bank revaluation processes.

Accounting Trigger in D365	Legal Entity	Type of Account	Ledger Account	Account Name	Debit	Credit	Accounting Rule	Logic for Debit/Credit
General Ledger Revaluation	USMF	Ledger	300120	Paid in Capital		1492.39	Golden-1	Offset impact of revaluation
	USMF	Ledger	801400	Currency Adjustment Loss-Unrealized	1492.39		Golden-2	Debit the increase in the Expense

Foreign Currency Revaluation- Bank

This is new functionality in D365 and it is accessed from the following path

Path: Cash and Bank Management→ Periodic tasks→ Foreign Currency Revaluation

The interface and process are the same as GL revaluation. The major difference is on the accounting side. Any exchange rate difference will go directly in the 'realized gain/loss' account since the bank balance is a liquid account.

Accounting Trigger in D365	Legal Entity	Type of Account	Ledger Account	Account Name	Debit	Credit	Accounting Rule	Logic for Debit/Credit
Bank Balance Revaluation	USMF	Ledger	110150	Bank Account-GBP		348.98	Golden-1	Offset impact of revaluation
	USMF	Ledger	801300	Currency Adjustment Loss-Realized	348.98		Golden-2	Debit the increase in the Expense

Foreign Currency Revaluation- Customer or Vendor

Following is the path to run the revaluation process for customers. The parameters are also more or less the same as ledger revaluation.

Path: Accounts Receivable→ Periodic tasks→ Foreign Currency Revaluation

Let's understand it using an end-to-end scenario in three steps:

1. <u>Invoicing to the customer</u>

We raised an invoice to a foreign customer in EUR and the exchange rate was 1.3698USD/EUR on that day. Following is the expected voucher entry:

Accounting Trigger in D365	Legal Entity	Type of Account	Ledger Account	Account Name	Debit (Euro)	Credit (Euro)	Amount (USD)	Accounting Rule	Logic for Debit/Credit
Invoicing on 31-May-2021 (1 Euro = 1.3698 USD)	USMF	Ledger	130110	Accounts Receivable-Foreign	10000		13698.63	Golden-1	Offset impact of revaluation
	USMF	Ledger	403700	Sales Account		10000	13698.63	Golden-2	Credit the increase in the Revenue/Gain

2. Revaluation at the month-end

Come month-end closing, revaluation of all foreign currency balances will be a step before you run closing. The exchange rate changes to 1.4285USD/EUR, so there will be exchange rate profit but it will be unrealized profit as the money is realized or collected from the customer yet.

Accounting Trigger in D365	Legal Entity	Type of Account	Ledger Account	Account Name	Debit (Euro)	Credit (Euro)	Amount (USD)	Accounting Rule	Logic for Debit/Credit
AR Revaluation on 30-June-21 (1 Euro = 1.4285 USD)	USMF	Ledger	130110	Accounts Receivable- Foreign			587.08	Golden-1	Offset impact of revaluation
	USMF	Ledger	801600	Currency Adjustment Profits- Unrealized			(587.08)	Golden-2	Credit the increase in the Revenue/Gain

3. Customer payment next month

The customer makes the payment next month and the rate has further increased to 1.5384USD/EUR, so there will be more exchange rate profit. All the gain/loss calculated previously will be reversed automatically ($587.08 in this example) and a new amount will be calculated with the current rate which will be posted to realized profit account. ($1,685.99 in this example).

Accounting Trigger in D365	Legal Entity	Type of Account	Ledger Account	Account Name	Debit (Euro)	Credit (Euro)	Amount (USD)	Accounting Rule	Logic for Debit/Credit
	USMF	Ledger	110130	Bank Account- EUR	10,000.00			Golden-1	Debit the increase in the asset
	USMF	Ledger	130110	Accounts Receivable- Foreign		10,000.00		Golden-1	Credit the decrease in the asset
Customer pays on 5-July-21 (1 Euro = 1.5384 USD)	USMF	Ledger	130110	Accounts Receivable- Foreign			(587.08)		Reversal of previous revaluation of AR
	USMF	Ledger	801600	Currency Adjustment Profits- Unrealized			587.08		
	USMF	Ledger	130110	Accounts Receivable- Foreign			1,685.99		Offset impact of revaluation
	USMF	Ledger	801500	Currency Adjustment Profits- Realized			#######	Golden-2	Credit the increase in the revenue/profit

We will not discuss the vendor balance revaluation as it is just based on the same concept as we discussed above with the difference of transaction direction.

6.4 Inter-Unit Accounting

This feature is one of my favourites in D365. Many of our clients had a requirement to have a segment-wise balance sheet. The segment can be a business unit, department, Production site, and so on. It was a common requirement from a Profit & Loss perspective but maintaining balance sheets for each segment comes with a lot of overhead. For example- accountants must make sure that they have entered the same business unit dimension in the debit line and credit line while posting a transaction.

I. Balancing financial dimension

Define the dimension which is required for the balance sheet. In the example below, 'business unit' has been set up as a balancing dimension. *Please note, you can choose only one dimension to get the balance sheet. If the management wants to get a balance sheet based on business unit and department, it will not be possible.*

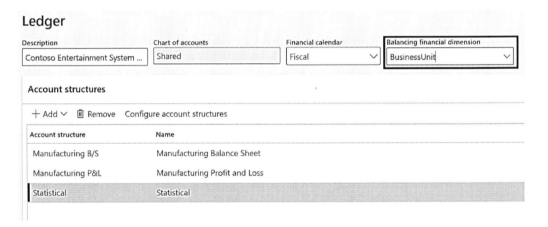

II. Define the debit/credit accounts for balancing.

You might need to create these two new accounts in your chart of accounts if you don't have these already. They will be the balancing account or you can say that they will be used to complete an accounting entry at the segment level.

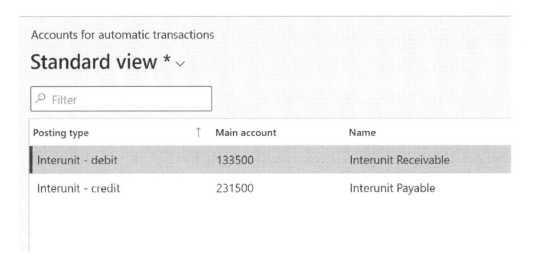

Accounts for automatic transactions

Standard view * ∨

Posting type ↑	Main account	Name
Interunit - debit	133500	Interunit Receivable
Interunit - credit	231500	Interunit Payable

III.   Another important thing to note is that when you enable a particular segment as a 'balancing financial dimension' in the ledger, you must make sure that this dimension should be mandatory for every transaction. Technically, you can make sure of this control using the 'configure account structure' in general ledger setup.

See the example below, there are two account structures in this entity, and both have 'business unit' as a mandatory dimension to enter.

Account structures | RETAIL B/S - US : RETAIL BALANCE SHEET - UNITED STATES

Retail B/S - US

Segments and allowed values

+ Add   Remove   Duplicate   Validate   Add segment   Segment actions ∨

☐ Only show overlapping rows

✓	MainAccount ↑	BusinessUnit	Department
	100000..399999	*	""..*

Account structures | RETAIL P&L - US : RETAIL PROFIT & LOSS - UNITED STATES

Retail P&L - US

Segments and allowed values

+ Add   🗑 Remove   🗋 Duplicate   Validate   Add segment   Segment actions ∨

☐ Only show overlapping rows

✓	MainAccount ↑	BusinessUnit	Department	CostCenter
	400000..999999	005..006;078;*	"";022..027;033	"";015..021;086

Technically, you just must make sure that 'allow blank value' is not checked for any of the account structures for that segment.

Let's see this inter-unit accounting now in action after we have configured everything required.

We entered a voucher with two different business units (e.g., 002 and 001 in the example). When you post the voucher the debit will go to 002 and credit will go to 001 which makes creates an imbalance at the unit level.

Journal voucher

But as we have configured the inter-unit accounting and business unit is defined as balancing dimension, any imbalance at the BU level will be balanced by the newly created accounts. See the below voucher after posting in D365, there are two extra lines which just to balance.

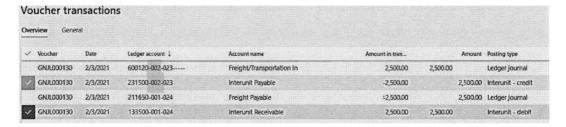

Voucher transactions

✓	Voucher	Date	Ledger account ↓	Account name	Amount in tran...	Amount	Posting type
	GNJL000130	2/3/2021	600120-002-023------	Freight/Transportation In	2,500.00	2,500.00	Ledger journal
✓	GNJL000130	2/3/2021	231500-002-023	Interunit Payable	-2,500.00	2,500.00	Interunit - credit
	GNJL000130	2/3/2021	211650-001-024	Freight Payable	-2,500.00	2,500.00	Ledger journal
✓	GNJL000130	2/3/2021	133500-001-024	Interunit Receivable	2,500.00	2,500.00	Interunit - debit

To make it easier, see the same data in the below grid. Set-1 represents the set of lines that are balanced at 002 level. Debit is under 'Freight expense' and credit is under the 'Interunit payable' account. If you just see the balance sheet of '002' business unit, you will see that there is a liability amounting to $2,500 under the 'interunit payable' account.

Similarly, if you analyze the balance sheet of '001' business unit, you will see an asset of $2,500 in the name of 'interunit receivable.'

Accounting Trigger in D365	Type of Account	Account	Ledger Account	Business Unit (Dimension-1)	Dimension-2 (Department)	Debit	Credit	
Manual	Ledger	Freight Expense	600120	002	023	2500		Set-1
Automatic	Ledger	Interunit payable	231500	002	023		2500	
Manual	Ledger	Freight Payable	211650	001	024		2500	Set-2
Automatic	Ledger	Interunit receivable	133500	001	024	2500		

7

Procurement to Pay in D365 ERP

Procure to pay (purchase to pay or P2P) is the process of obtaining and managing the raw materials needed for manufacturing a product or providing a service, receiving these materials, and paying to the vendor. For trading organizations as well, procurement is a basic process where the products are bought from the vendor, received in warehouses after quality inspections, and then the suppliers are paid based on the invoices.

There are related financial documents needed to carry out these transactions.

In any organization, procurement can be of two types, direct and indirect. While direct procurement deals with the products and services which are related to direct revenue-generating mechanisms for the organization, indirect procurement is related to internal organizational requisitions for services and equipment. An example of a direct procurement can be for a juice manufacturing company to procure fruits as the raw materials. Whereas within the same organization there can be requests for procuring computers, hardware, or services related to the plant, machinery, and any other miscellaneous products and services which are not directly contributing to the revenue generation process of the organization. Such procurement is termed indirect procurement.

The related transactions in Dynamics 365 Finance where we encounter financial entries in the Procure to pay cycle are the following:

1. Receipt of Goods/Services

2. Invoice Posting

3. Process Payment

Below is a high-level process flow for procure to pay and the processes wherever there is an accounting impact.

Fig. 7.1

As we identified in the flow above (Fig. 7.1), the first point of accounting entries in D365 ERP is at the point of "Receipt". Let us look at some of the samples of accounting entries for receipt.

Accounting for Stock Items/Products

Product Receipt

As per Fig 7.1, the first process where financial accounting occurs for the Procure to pay cycle is the Receipt of Goods or Services. This process is the actual receipt of goods or services in the warehouse of the company based on the purchase order.

Accounting for Product Receipt

Stocked products from the name itself are items that the company physically maintains or stocks in inventory. The following set of entries happen when stocked products (goods that are being stocked in inventory) are received.

Accounting Trigger in D365	Account Family	Type of Account	Account	Sub ledger impacted	Transaction type in D365	Debit	Credit	Accounting Rule	Logic for Debit/Credit
Product Receipt	Inventory	Asset	Inventory A/c*	Yes	Cost of purchased materials received	X		Golden-1	Debit the increase in the Asset
	Ledger	Asset	Raw Materials Receipts	No	Purchase expenditure, un-invoiced		X	Golden-1	WIP A/c
	Ledger	Asset	Raw Materials Receipts	No	Purchase expenditure, un-invoiced	X		Golden-2	WIP A/c
	Creditor	Liability	Accrued Purchases - Received Not Invoiced	No	Accrued Purchases - Received Not Invoiced		X	Golden-1	Credit the increase in the liability

Notes-

1. * As per my experience, some companies prefer to use a control account during this process instead of a final inventory account. This separates the physical ledger postings from the financial ledger postings for purchase order transactions which is very helpful during inventory reconciliation. This level of segregation would ease the burden during Inventory to Ledger reconciliation.

2. 'Transaction type in D365' above represents the posting profile type.

3. The first two lines in the above voucher are always there for all stocked products. The last two lines (in blue) are optional and depend on the flag 'Accrue liability on product receipt' in the item model group. It will purely be driven by the company policy if the company wants to recognize any liability upon product receipt.

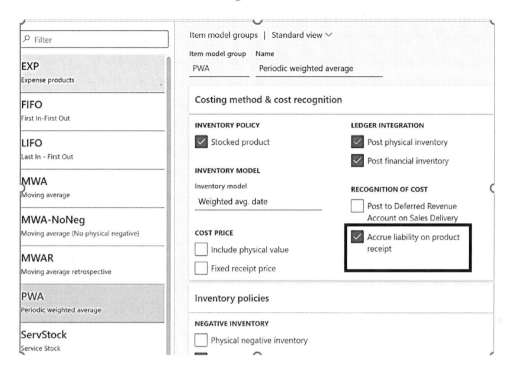

4. 'Purchase Expenditure-Un-invoiced' will be debited and credited with the same amount, so it will not be visible in the product receipt voucher. (For the scenario mentioned in the above point)

5. Irrespective of the above flag, all the voucher lines are reversed when the purchase order is invoiced, and a new voucher is generated. So, what happens during product receipt will not make much difference if there is not much time difference between product receipt and invoicing steps.

Here is a sample voucher entry for stocked products receipt.

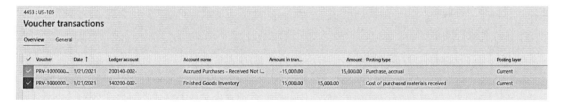

The accounts that would be configured for this purpose are present under *Cost management -> Ledger Integration policies setup-> Posting -> Purchase Order. See below picture for accounts that are required to be configured.*

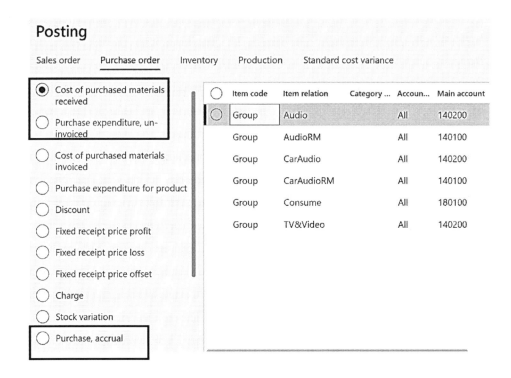

Accounting for Product Receipt with Standard Costing

The accounting that is discussed above for product receipt is in the case, where the costing method is other than Standard cost. In the case of standard cost, there is an additional accounting entry – *Purchase price variance or in terms of accounting known as PPV*. This is set up under the *Standard Cost variance* tab of Ledger postings under *Cost Management -> Ledger integration policies setup -> Posting -> Standard cost variance.*

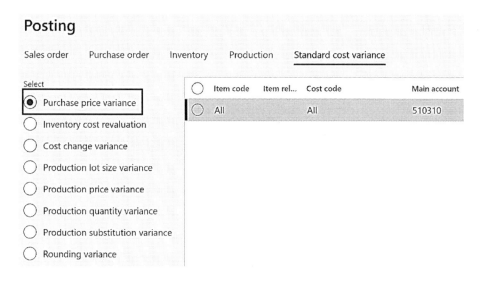

Let us understand what Standard cost is and what is the relevance of Purchase price variance.

Here is an example of how the Purchase price variance works. The standard cost on the released product/item is set to $950 as below.

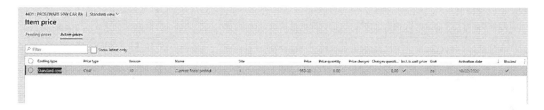

The unit price on the Purchase order is set to $1000, which means that you are about to buy a product that as per your calculations is supposed to be $950 at $1000. The $50 difference would be posted Purchase price variance as a loss, as you are spending more than what you have estimated/established the cost to be.

Below is the accounting entry for a product/item that is valued at standard cost.

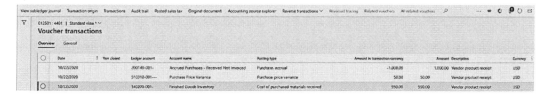

Accounting Trigger in D365	Account Family	Type of Account	Account	Sub ledger impacted	Transaction type in D365	Debit	Credit	Accounting Rule	Logic for Debit/Credit
Product Receipt with Std. Costing	Inventory	Asset	Inventory A/c*	Yes	Cost of purchased materials received	950		Golden-1	Debit the increase in the Asset
	Ledger	P&L	Purchase Price Variance	No	Purchase Price Variance	50		Golden-2	Balancing entry *
	Ledger	Asset	Raw Materials Receipts	No	Purchase expenditure, un-invoiced		1000	Golden-1	WIP A/c

Notes

- * Variance account is taken as a balancing figure, if the purchase price is more than standard cost, then the variance is debited to make the voucher balanced.

- The purchase price variance is set to be Profit and Loss account.

Definition of Standard Cost

A standard cost is described as a predetermined cost, an estimated future cost, an expected cost, a budgeted unit cost, a forecast cost, or as the "should be" cost. Standard costs are often an integral part of a manufacturer's annual profit plan and operating budgets.

When standard costs are used in a manufacturing setting, a product's standard cost for a future accounting period will consist of the following:

- *Direct materials: a standard quantity of each material and a standard cost per unit of material*

- *Direct labor: a standard quantity of labor and a standard cost per hour of labor*

- *Manufacturing overhead: a budget for the fixed overhead, the standard variable overhead rate, and the standard quantity for applying a fixed and variable overhead rate*

In a standard costing system, the standard costs of the manufacturing activities will be recorded in the inventories and the cost of goods sold accounts. Since the company must pay its vendors and production workers the actual costs incurred, there are likely to be some differences. The differences between the standard costs and the actual manufacturing costs are referred to as cost variances and will be recorded in separate variance accounts. Any balance in a variance account indicates that the company is deviating from the amounts in its profit plan.

While standard costs can be a useful management tool for a manufacturer, the manufacturer's external financial statements must comply with the cost principle and the matching principle. Therefore, significant variances must be reviewed and properly assigned or allocated to the cost of goods sold and/or inventories.

Invoice Posting

As per Fig 7.1, Invoice posting is another process in procure to pay cycle where financial transactions are created. Vendor invoice posting is the process that records the company's liabilities to the vendor. This results in an increase in the open vendor balance.

Invoicing through the purchase order is done for all 'product-related purchases, whether the product is an item-type or service-type item.

Accounting on invoicing of stocked products
Below shows the accounting that is created when the product receipt for a stocked item is invoiced.

Accounting Trigger in D365	Account Family	Type of Account	Account	Sub ledger impacted	Transaction type in D365	Debit	Credit	Accounting Rule	Logic for Debit/Credit
Invoicing	Inventory	Asset	Inventory A/c	Yes	Cost of purchase material Invoiced	X		Golden-1	Debit the increase in the Asset
	Creditor	Liability	Accounts Payable	Yes			X	Golden-1	Credit the increase in the liability
	Inventory	Asset	Inventory A/c	Yes	Cost of purchased materials received		X	Golden-1	Reversal of the Product receipt voucher
	Creditor	Liability	Accrued Purchases - Received Not Invoiced	No	Accrued Purchases - Received Not Invoiced	X		Golden-1	

In this accounting entry, the first two lines are the final accounting entries that would happen upon posting the invoice. The last two rows are the reversal of the product receipt voucher.

Here is a sample voucher entry from D365 for the stocked products invoice.

When a purchase order is product receipt and vendor invoice updated for stocked items, there are two accounting processes and resulting entries which occur:

- The accrued liability is updated (this new feature was released in AX 2012)

- The inventory cost is updated (this new feature was released in AX 2009)

Two new posting types were made available to support these transactions.

Document ↑	Subledger jour...	Posting type ↑	Ledger account	Transaction cur...	Debit	Credit	Accounting cur...	Accounting cur...	Reporting curr...	Reporting c
PIV-110000993	06425	Vendor balance	200100-002-	USD	0.00	15,000.00	0.00	15,000.00	0.00	15,0
PIV-110000993	06425	Purchase expenditure for product	600180-002---Audio	USD	15,000.00	0.00	15,000.00	0.00	15,000.00	
PIV-110000993	06425	Purchase expenditure for product	600180-002---Audio	USD	0.00	15,000.00	0.00	15,000.00	0.00	15,0
PIV-110000993	06425	Cost of purchased materials received	140200-002-	USD	0.00	15,000.00	0.00	15,000.00	0.00	15,0
PIV-110000993	06425	Purchase expenditure, un-invoiced	600180-002---Audio	USD	15,000.00	0.00	15,000.00	0.00	15,000.00	
PIV-110000993	06425	Purchase expenditure, un-invoiced	600180-002---Audio	USD	0.00	15,000.00	0.00	15,000.00	0.00	15,0
PIV-110000993	06425	Cost of purchased materials invoiced	140200-002-	USD	15,000.00	0.00	15,000.00	0.00	15,000.00	
PIV-110000993	06425	Purchase, accrual	200140-002-	USD	15,000.00	0.00	15,000.00	0.00	15,000.00	

Purchase expenditure, un-invoiced: This posting type is used to accrue the liability and product receipt amounts when a purchase order is product receipt updated.

For the posting type 'Purchase expenditure, un-invoiced', those accounting entries which net to 0 will not be transferred to the general journal. These postings are not displayed in the Voucher transactions form but in the Subledger journal form. The Subledger journal form will display all generated subledger accounting entries, including those not transferred to the general ledger.

Purchase expenditure for product: This posting type is used to offset the liability and inventory amounts when a purchase order invoice is updated.

For posting type Purchase expenditure for product, those accounting entries that net to 0 will no longer display both a line for the credit and the debit entries in the Voucher transactions form but will be displayed in the Subledger journal and be transferred to the general ledger in a single zero-amount accounting entry.

The accounts that would be configured for this purpose are present under Cost *management -> Ledger Integration accounting policies-> Posting -> Purchase Order. See the below picture for accounts that are required to be configured.*

There will be scenario for both posting types when there will be a balance for stocked items e.g. such as from currency translation, or for a stocked item which is returned with a variance between the unit price and the average cost.

'Purchase Expenditure for Product' is effectively the clearing account between accounts payable and inventory. This account is also used to book a difference in original cost and the amount a vendor allows for the return when using a costing method other than standard costing.

Posting

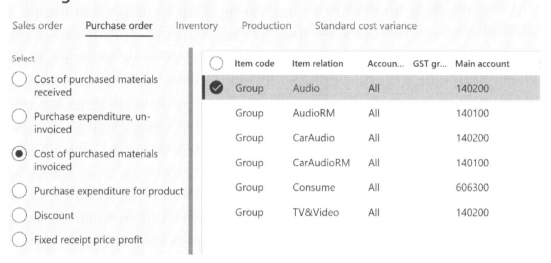

Posting

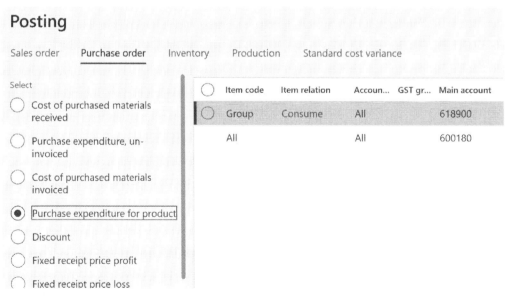

Accounting for non-stocked products

Below are the accounting entries that would occur at the point of invoicing for any non-stocked products or categories are invoiced.

Product Receipt (Accrual of Expenses/liabilities)

Generally, non-stock products (service items) don't need any formal receiving in the purchase order but taking a conservative approach, many organizations like to do that and accrue the liability and expense for the services.

Accounting Trigger in D365	Account Family	Type of Account	Account	Sub ledger impacted	Transaction type in D365	Debit	Credit	Accounting Rule	Logic for Debit/Credit
Product Receipt	Ledger	Expense	Raw Materials Receipts	No	Purchase expenditure, un-invoiced	X		Golden-2	Debit the increase in the expense
	Creditor	Liability	Accrued Purchases - Received Not Invoiced	No	Accrued Purchases - Received Not Invoiced		X	Golden-1	Credit the increase in the liability

Invoicing of 'Non-Stock' products or Services

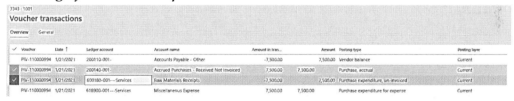

Accounting Trigger in D365	Account Family	Type of Account	Account	Sub ledger impacted	Transaction type in D365	Debit	Credit	Accounting Rule	Logic for Debit/Credit
Non Stock PO Invoicing	Ledger	Expense	Miscellaneous Expense	No	Purchase Expenditure for Expense	X		Golden-2	Debit the increase in the expense
	Creditor	Liability	Accounts Payable	Yes	Vendor		X	Golden-1	Credit the increase in the liability
	Ledger	Expense	Raw Materials Receipts	No	Purchase expenditure, un-invoiced		X	Golden-2	Reversal of the Product Receipt Voucher
	Creditor	Liability	Accrued Purchases - Received Not Invoiced	No	Accrued Purchases - Received Not Invoiced	X		Golden-1	

The account setup for Purchase expenditure for the expense is present under *Cost management -> Ledger Integration accounting policies-> Posting -> Purchase Order*

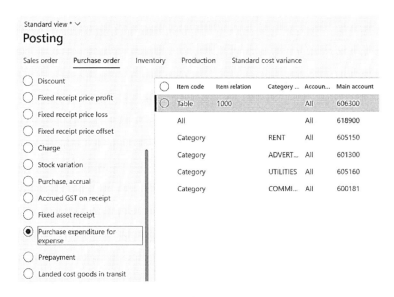

The account setup for Vendor balance is present under *Accounts Payable -> Setup -> Vendor posting profiles -> Summary account.*

Invoicing through Vendor Invoice Journal.

Invoicing through the vendor invoice journal is done for 'non-item related' transactions. Any vendor invoice that is not posted through purchase order can be invoiced through the invoice journal.

Accounting on invoice journal posting

Below are the accounting entries that are generated when a vendor invoice journal is posted:

Accounting Trigger in D365	Account Family	Type of Account	Account	Sub ledger impacted	Transaction type in D365	Debit	Credit	Accounting Rule	Logic for Debit/Credit
Non Stock PO Invoicing	Ledger	Expense	Miscellaneous Expense	No		X		Golden-2	Debit the increase in the expense
	Creditor	Liability	Accounts Payable	Yes	Vendor		X	Golden-1	Credit the increase in the liability

Prepayment against a Purchase Order

The business needs to give prepayment to the vendors against the purchase order. Generally, the advance amount is a certain percentage of the order amount. There is a feature in D365 to define this percentage and process an invoice for this prepayment. This prepaid amount can be settled finally when the invoice is paid.

Accounting on invoice journal posting

Below are the accounting entries that are generated when a vendor invoice is posted for the prepayment:

Accounting Trigger in D365	Account Family	Type of Account	Account	Sub ledger impacted	Transaction type in D365	Debit	Credit	Accounting Rule	Logic for Debit/Credit
Vendor Prepayment Invoice	Ledger	Asset	Prepaid A/c	No	Prepayment	X		Golden-1	Debit the increase in the asset
	Vendor	Liability	Accounts Payable	Yes	Vendor Balance		X	Golden-1	Credit the increase in the liability

Sample transaction from D365 with actual ledger accounts.

Voucher transactions

Overview   General

✓	Voucher	Date	Ledger account ↑	Account name	Amount in tran...		Amount	Posting type	Posting layer	Transaction type
	PPP-140000	2/1/2021	132190--	Prepaid Other Expenses	15,267.92	20,914.96		Prepayment	Current	Purchase order
	PPP-140000	2/1/2021	200100--	Accounts Payable - Domestic	-15,267.92		20,914.96	Vendor balance	Current	Purchase order

Please note that this transaction creates a credit for the supplier and the supplier is paid with another transaction which is a payment transaction. Processing a payment is the process that is under the accounts payable module→payment journal. It is a common process to pay all types of invoices. (See the next section)

Process Payment

As per Fig 7.1, vendor payment processing is another point in procure to pay cycle where financial transactions are created. The vendor payment process reduces the company's liability to vendors.

Accounting on vendor payment processing

Below are the accounting entries that are generated when a vendor payment journal is posted:

Accounting Trigger in D365	Account Family	Type of Account	Account	Sub ledger impacted	Transaction type in D365	Debit	Credit	Accounting Rule	Logic for Debit/Credit
Vendor Payment	Vendor	Liability	Accounts Payable	Yes	Vendor Balance	X		Golden-1	Debit the decrease in the liability
	Bank	Asset	Bank A/c	Yes	Bank		X	Golden-1	Credit the decrease in the asset

Important-

There are a few more posting types under purchase tab in inventory posting profile e.g.

- Fixed Receipt Price Profit

- Fixed Receipt Price Loss

- Fixed Receipt Price Offset

We will be covering those in **Chapter 10A** 'Inventory Valuation Methods' even though these types are very rarely used now after standard costing is introduced.

8

Fixed Assets in D365 ERP

Asset Acquisition

A fixed asset can be acquired or capitalized from one of the following processes in Microsoft D365:

- Purchase Order Process

- Invoice Journal/Fixed Asset Acquisition Journal

- Internal Project (Capital in Process)

- Convert Inventory into Fixed Asset

We will be talking about the first two ways of acquisition in this book.

Asset Acquisition with Purchase Order Process

Product Receipt

As we discussed earlier during the inventory purchase order discussion in Fig. 7.1, the first process where financial accounting occurs for the Procure to pay cycle is the Receipt of Goods or Services. This process is the actual receipt of goods or services in the warehouse of the company based on the purchase order, but some organizations choose not to have the receiving step for fixed asset items. So, this step can be optional for them. (I am not detailing the accounting here again as it will be the same as any other purchase)

Invoice Posting

Invoicing voucher for a fixed asset depends on the following flag on the 'Fixed Assets' parameter form which is called 'Allow asset acquisition from purchasing'.

- **If this flag is set to 'Yes',** it means that the asset will be capitalized when the purchase order is posted.

- On the other hand, **if this flag is set to 'No',** it means that the purchase order invoice will park the asset to a ledger account and the asset will be capitalized through an acquisition journal entry.

Fixed assets parameters

Fixed assets	Set up fixed asset information
Number sequences	, ,
Maintenance	**Accounting rules**

Accounting rules

Barcode

Purchase orders

Allow asset acquisition from Purcha... ⬤ Yes

Create asset during product receipt ... ⬤ Yes

Restrict asset acquisition posting to us... ⌄

Check for fixed assets creation durin... ⬤ Yes

Let's understand this in detail with sample vouchers-

'Allow Acquisition from purchasing' is YES
Below shows the accounting that is created when the product receipt for fixed asset is invoiced.

Accounting Trigger in D365	Account Family	Type of Account	Account	Sub ledger impacted	Transaction type in D365	Debit	Credit	Accounting Rule	Logic for Debit/Credit	
Purchase Order Invoicing	Inventory	Asset	Tangible Fixed Asset	Yes	Cost of purchased materials received		X	Golden-1	Reversal of the Product receipt voucher	Step-1
	Creditor	Liability	Accrued Purchases - Received Not Invoiced	No	Accrued Purchases - Received Not Invoiced	X		Golden-1		
	Fixed Asset	Asset	Tangible Fixed Asset	No	Fixed Asset Receipt	X		Golden-1	Debit the increase in the Asset	Step-2
	Creditor	Liability	Accounts Payable	Yes	Vendor balance		X	Golden-1	Credit the increase in the liability	
	Fixed Asset	Asset	Tangible Fixed Asset	Yes	Fixed assets, debit	X		Golden-1	Debit the increase in the Asset	Step-3
	Fixed Asset	Asset	Tangible Fixed Asset	No	Fixed asset issue		X	Golden-1	Credit the decrease in the asset	

Let's understand this voucher in the following steps:

Step-1: First two lines are the exact reversal of the product receipt voucher.

Step-2: Next two lines are the normal invoicing voucher where the fixed asset account is debited, and the supplier account is credited.

Step-3: It is the *capitalization voucher* where fixed asset sub-ledger is debited, and fixed asset ledger is credited.

The accounts that would be configured for this purpose are present under Cost *management -> Ledger Integration accounting policies-> Posting -> Purchase Order. See the below picture for accounts that are required to be configured.*

'Fixed Asset, receipt' account from step-2 above comes from the 'fixed asset receipt' type under 'purchase order' in the Item Posting profile.

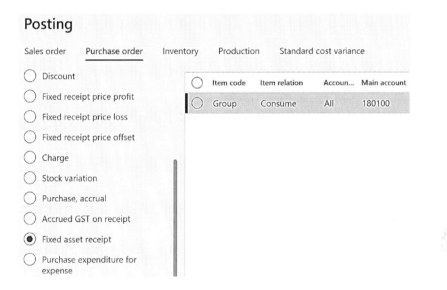

Fixed Asset, debit account from step-3 above comes from the 'Acquisition' ledger under Fixed Assets posting profile.

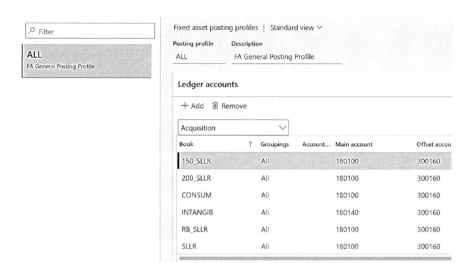

'Fixed Asset, issue' account from step-3 above comes from the 'fixed asset issue' type under inventory in Item Posting profile.

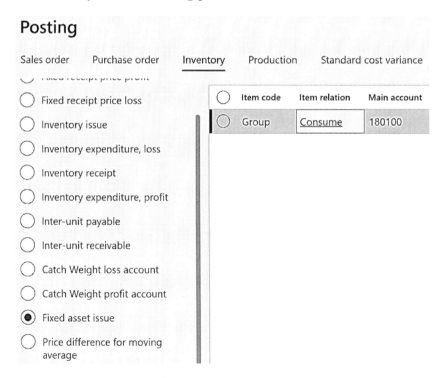

Here is a sample voucher entry from D365 for the stocked-products invoice.

Voucher transactions

Overview | General

✓	Voucher	Date ↑	Ledger account	Account name	Amount in tran...		Amount	Posting type	Posting... ∇	Transaction type
✓	PIV-110001000	2/2/2021	180100-001-	Tangible Fixed Assets	17,000.00	17,000.00		Fixed assets, debit	Current	Purchase order
	PIV-110001000	2/2/2021	180100-001-	Tangible Fixed Assets	17,000.00	17,000.00		Cost of purchased materials invoiced	Current	Purchase order
	PIV-110001000	2/2/2021	180100-001-	Tangible Fixed Assets	-17,000.00		17,000.00	Cost of purchased materials received	Current	Purchase order
✓	PIV-110001000	2/2/2021	180100-001-	Tangible Fixed Assets	-17,000.00		17,000.00	Fixed asset issue	Current	Purchase order
	PIV-110001000	2/2/2021	618900-001---	Miscellaneous Expense	0.00	0.00		Purchase expenditure for product	Current	Purchase order
	PIV-110001000	2/2/2021	200110-001-	Accounts Payable - Other	-17,000.00		17,000.00	Vendor balance	Current	Purchase order
	PIV-110001000	2/2/2021	200140-001-	Accrued Purchases - Received Not Invoiced	17,000.00	17,000.00		Purchase, accrual	Current	Purchase order

'Allow Acquisition from purchasing' is NO

In this option as well, the accounting entries are the same except for the step-3 voucher. In other words, the capitalization of the asset (step-3) takes place through a separate journal.

Following is the accounting voucher upon invoicing of the purchase order.

Accounting Trigger in D365	Account Family	Type of Account	Account	Sub ledger impacted	Transaction type in D365	Debit	Credit	Accounting Rule	Logic for Debit/Credit	
Purchase Order Invoicing	Inventory	Asset	Tangible Fixed Asset	Yes	Cost of purchased materials received		X	Golden-1	Reversal of the Product receipt voucher	Step-1
	Creditor	Liability	Accrued Purchases - Received Not Invoiced	No	Accrued Purchases - Received Not Invoiced	X		Golden-1		
	Fixed Asset	Asset	Tangible Fixed Asset	Yes	Fixed Asset Receipt	X		Golden-1	Debit the increase in the Asset	Step-2
	Creditor	Liability	Accounts Payable	Yes	Vendor balance		X	Golden-1	Credit the increase in the liability	

Following is the sample voucher for this scenario:

Voucher transactions

Overview | General

✓	Voucher	Date ↑	Ledger account	Account name	Amount in tran...		Amount	Posting type	Posting layer	Transaction type
	PIV-110000997	2/1/2021	180140-001-	Intangible Fixed Assets	18,000.00	18,000.00		Cost of purchased materials invoiced	Current	Purchase order
	PIV-110000997	2/1/2021	180100-001-	Tangible Fixed Assets	-18,000.00		18,000.00	Cost of purchased materials received	Current	Purchase order
	PIV-110000997	2/1/2021	618900-001---	Miscellaneous Expense	0.00	0.00		Purchase expenditure for product	Current	Purchase order
	PIV-110000997	2/1/2021	200110-001-	Accounts Payable - Other	-18,000.00		18,000.00	Vendor balance	Current	Purchase order
	PIV-110000997	2/1/2021	200140-001-	Accrued Purchases - Received Not Invoiced	18,000.00	18,000.00		Purchase, accrual	Current	Purchase order

Following is the voucher when the asset is capitalized using an acquisition journal.

Accounting Trigger in D365	Account Family	Type of Account	Account	Sub ledger impacted	Transaction type in D365	Debit	Credit	Accounting Rule	Logic for Debit/Credit	
Acquisition proposal	Fixed Asset	Asset	Tangible Fixed Asset	Yes	Fixed assets, debit	X		Golden-1	Debit the increase in the Asset	Step-3
	Fixed Asset	Asset	Tangible Fixed Asset	No	Fixed asset issue		X	Golden-1	Credit the decrease in the asset	

Sample of an acquisition journal below.

Path: Fixed Assets→Journal Entries→ Create acquisition proposal

Sample of an acquisition journal below.

Acquisition proposal

Parameters

JOURNAL

Posting layer

Current

Name of journal

FACur

Records to include

▽ Filter

FIXED ASSETS

Fixed asset number

Fixed asset group

FIXED ASSET BOOK

Book

Status

Not yet acquired

Run in the background

Create journal

When this proposal is created, a journal is created on the following path.

Path: Fixed Assets→Journal Entries→ Fixed Asset Journal

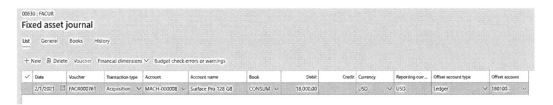

Post this journal and check the voucher which will be something like this.

✓	Voucher	Date ↑	Ledger account	Account name	Amount in tran...		Amount	Posting type	Posting... ▽	Transaction type
	FACR000761	2/1/2021	180100--	Tangible Fixed Assets	18,000.00	18,000.00		Fixed assets, debit	Current	Fixed assets
	FACR000761	2/1/2021	180100--	Tangible Fixed Assets	-18,000.00		18,000.00	Ledger journal	Current	Fixed assets

Voucher transactions — Overview / General

Asset Acquisition with Journals (Non-Purchase Order acquisition)

Small organizations generally don't have a full-fledged procurement process, so they don't raise the purchase order for every asset purchase. Even some larger organizations don't follow the process for assets of insignificant value. In that case, the whole acquisition process is very quick and can be achieved through the journals e.g. 'Fixed Asset Journal' or 'Vendor Invoice Journal'

Following is the voucher when the asset is acquired using a Vendor Invoice journal.

Accounting Trigger in D365	Account Family	Type of Account	Account	Sub ledger impacted	Transaction type in D365	Debit	Credit	Accounting Rule	Logic for Debit/Credit
Vendor Invoice Journal	Fixed Asset	Asset	Tangible Fixed Asset	Yes	Fixed assets, debit	X		Golden-1	Debit the increase in the Asset
	Fixed Asset	Liability	Accounts Payable	Yes	Vendor balance		X	Golden-1	Credit the increase in the liability

Acquisition Adjustment

Accounting treatment of acquisition adjustment is very much similar to the acquisition, but it has its posting profile set up. Some organizations want to track all adjustments in a separate account, so this is the flexibility comes from a separate type of posting profile i.e., Acquisition Adjustment.

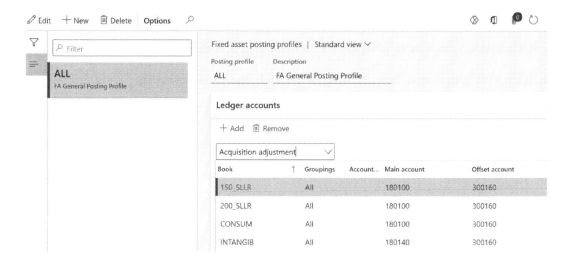

Just an important point to note about adjustment is that there is a configuration required to be done before you can make any adjustment in the existing asset. Select the "Allow Multiple Acquisition" checkbox in a fixed asset's parameters.

Path: Fixed Asset→Setup→Fixed Asset Parameters→allow multiple acquisitions.

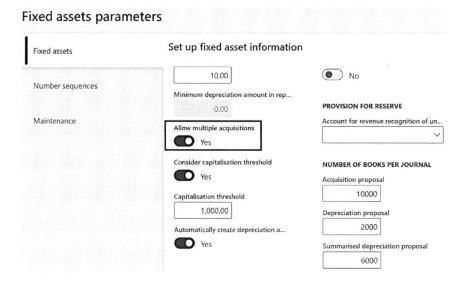

Depreciation/Depreciation Adjustment

Fixed asset depreciation is a fully automated process if the configuration is well defined. It works by calculating and generating a fixed asset's deprecation according to the associated depreciation method in the asset book.

Following is the voucher when the depreciation process is run through the depreciation proposal process or manually for an individual asset.

Accounting Trigger in D365	Account Family	Type of Account	Account	Sub ledger impacted	Transaction type in D365	Debit	Credit	Accounting Rule	Logic for Debit/Credit
Depreciation	Fixed Asset	P&L	Depreciation Expens	Yes	Ledger Journal	X		Golden-2	Debit the increase in the Expense
	Fixed Asset	Liability	Accumulated Depreciation	Yes	Fixed Assets, credit		X	Golden-1	Credit the decrease in the asset

Following is the posting profile for depreciation. Please note that the offset account will be the expense account and the main account represents the asset account.

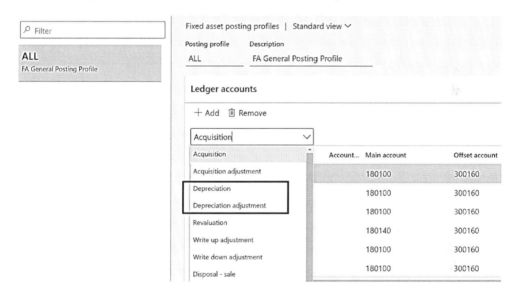

Revaluation/Write Up/down Adjustment.

- Revaluation – A change in the value of an asset. Revaluation can be used for both write-up and write-down transactions but is a separate transaction type because revaluation proposals are available in the Fixed assets journal.

- Write-down adjustment – A downward adjustment in the accounting value of an asset. The offset account is usually a profit and loss account. This is the most common adjustment.

- Write-up adjustment – An upward adjustment in the accounting value of an asset. The offset account is usually a different balance sheet account. This is a less common adjustment.

Disposal/Sale/Scrap

This is the most complicated accounting and posting profile of D365. Believe me, it has many layers of accounting and posting, and the local regulations of each country might make it more difficult to understand sometimes.

In general, when you sell any asset, everything which is posted in previous years and current year (e.g. acquisition and accumulated depreciation, but not the expense side of depreciation) are reversed and the net value of both will be compared with the sale amount to calculate the profit or loss on that transaction e.g. you bought a laptop in 2019 for $2,000 and depreciated it to the extent of $800 in two years, then the net book value of the laptop will be $1,200 in 2021. If you sell it for $1,000 now, there will be a net loss of $200 ($1,200-$1,000).

The above calculation looks straightforward, now let's see how different posting accounts are defined for this transaction.

Example-

In the below example, the asset is sold for $1,000 and it was bought for $100,000 and there was a depreciation amount of $14,611.14 by that time. So, the net loss on the same will be $84,388.86.

Let's create a free text invoice to see the asset for $1,000. See the account which defaults on the free text invoice line when you enter the fixed asset number in the line. (It appears with 'Customer Revenue' as posting type in the voucher)

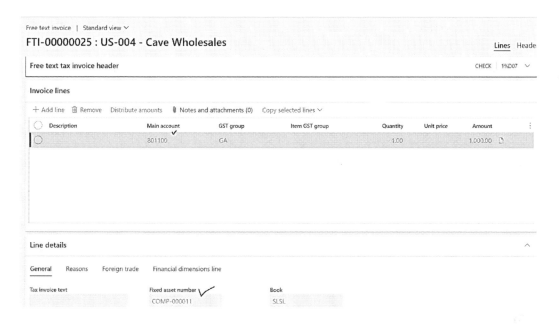

Following is the voucher entry that's generated upon posting the free text invoice.

Accounting Trigger in D365	Account Family	Ledger account	Type of Account	Account name	Sub ledger impacted	Posting type	Amount	Total	Accounting Rule	Logic for Debit/Credit
Sale	Customer	130100	Asset	Accounts Receivable - Domestic	Yes	Customer balance	1,000.00	1,000.00	Golden-1	Debit the increase in the asset.
	Fixed Asset	180200	Asset	Accumulated Depreciation - Tangible Fixed Assets	Yes	Fixed assets, debit	14,611.14	14,611.14	Golden-1	Debit the increase in the asset.
	Ledger	801100	P&L	Gain & Loss - Disposal of Assets	Yes	Customer revenue	(1,000.00)			
	Ledger	801100	P&L	Gain & Loss - Disposal of Assets	Yes	Fixed assets, credit	(14,611.14)	84,388.86	Golden-2	Debit the increase in the Expense
	Ledger	801100	P&L	Gain & Loss - Disposal of Assets	Yes	Fixed assets, debit	185,388.86			
	Ledger	801100	P&L	Gain & Loss - Disposal of Assets	Yes	Fixed assets, credit	(85,388.86)			
	Fixed Asset	180100	Asset	Tangible Fixed Assets	Yes	Fixed assets, credit	(100,000.00)	(100,000.00)	Golden-1	Credit the decrease in the asset.

Analysis-

1. Account receivable (AR) will be created with $1,000.

2. Acquisition amount (100,000) and accumulated depreciation (14,611.14) amounts will be reversed i.e., credited and debited respectively.

3. Highlighted are the lines which represent the total profit or loss in this transaction.

4. It is important to map the same account for all of these posting types in the fixed assets posting profile so that profit or loss is lumped into the same account. See the posting profile for sale below and check the table where we have configured 801100 for the same purpose.

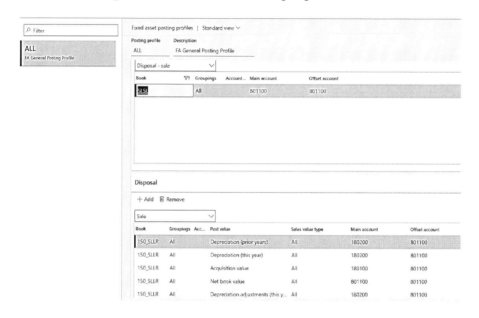

	Main Account	Offset Account	
Disposal Sale	801100	801100	This account appears in the free text invoice line level as a revenue account. Generally, this accounts' name is **'Gain & Loss- Disposal of Assets'** account. (See the image below)

Depreciation Prior year/Depreciation this year	180200	801100	The main Account is 'Accumulated depreciation' account and offset A/c is the Revenue or **'Gain & Loss- Disposal of Assets' account**
Acquisition Value/ Acquisition Adjustments	180100	801100	The main Account is 'Acquisition account' and offset A/c is the Revenue or **'Gain & Loss- Disposal of Assets' account**
Net Book Value	801100	801100	It will also be the same account **'Gain & Loss- Disposal of Assets'**

Scrap transaction also follows the similar posting profile. The only difference is that scrap doesn't carry any Accounts Receivable (AR) and revenue transactions, so a fixed asset journal is used to scrap the asset.

Accounting Entry-

Accounting treatment will also be same as sale except the two lines which are highlighted in yellow below for the obvious reasons.

Accounting Trigger in D365	Account Family	Ledger account	Type of Account	Account name	Sub ledger impacted	Posting type	Amount	Total	Accounting Rule	Logic for Debit/Credit
Sale	Customer	130100	Asset	Accounts Receivable - Domestic	Yes	Customer balance	1,000.00	1,000.00	Golden-1	Debit the increase in the asset.
	Fixed Asset	180200	Asset	Accumulated Depreciation - Tangible Fixed Assets	Yes	Fixed assets, debit	14,611.14	14,611.14	Golden-1	Debit the increase in the asset.
	Ledger	801100	P&L	Gain & Loss - Disposal of Assets	Yes	Customer revenue	(1,000.00)			
	Ledger	801100	P&L	Gain & Loss - Disposal of Assets	Yes	Fixed assets, credit	(14,611.14)	84,388.86	Golden-2	Debit the increase in the Expense
	Ledger	801100	P&L	Gain & Loss - Disposal of Assets	Yes	Fixed assets, debit	185,388.86			
	Ledger	801100	P&L	Gain & Loss - Disposal of Assets	Yes	Fixed assets, credit	(85,388.86)			
	Fixed Asset	180100	Asset	Tangible Fixed Assets	Yes	Fixed assets, credit	(100,000.00)	(100,000.00)	Golden-1	Credit the decrease in the asset.

9

Order to Cash in D365 ERP

Order to Cash is a business process that involves, fulfilling customer requests/orders for goods and/or services in exchange for cash or payments.

The related transactions in Dynamics 365 where we encounter financial entries in the Order to Cash cycle are the following:

- Shipment of goods
- Invoicing to the customer
- Process Payment/Collection of payments from customer

Below is a high-level process flow for 'Order to cash' and the processes wherever there is an accounting impact.

Fig. 9.1

As we identified in the flow above (Fig. 9.1), the first point of accounting entries in Microsoft D365 is at the point of "Shipment of goods and services". Let us look at some of the samples of accounting entries for shipment or 'Packing slip' as we may call it in Dynamics 365 F&O terms.

Note: Accounting entries for Packing Slip would only be posted when the Accounts Receipt parameter 'Post packing slip in the ledger is marked along with;

- *for stocked items' model group(s) 'Post physical inventory 'is marked under the model group.*

3. *Shipment of Goods or Services*

As per Fig 9.1, the first process where financial accounting occurs for the Order to Cash cycle is the Shipment of Goods or Services.

Accounting on delivery of stocked products

Stocked products from the name itself are items that the company physically maintains or stocks in inventory. The following set of entries talks about the subsequent accounting which happens when stocked products (goods that are being stocked in inventory) are shipped.

Here is a sample voucher entry for a packing slip.

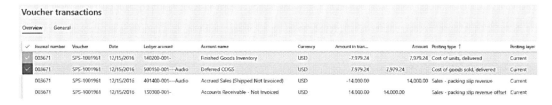

Accounting Trigger in D365	Account Family	Type of Account	Account	Sub ledger impacted	Transaction type in D365	Debit	Credit	Accounting Rule	Logic for Debit/Credit
Packing Slip	COGS	Expense	Deferred COGS	No	Cost of Goods Sold, delivered	X		Golden-2	Debit the increase in the expense
	Inventory	Asset	Finished Goods Inventory	Yes	Cost of Units, delivered		X	Golden-1	Credit the decrease in the asset (WIP)
	Sales	Revenue	Accrued Sales (Shipped Not invoiced)	No	Sales-Packing Slip Revenue		X	Golden-2	Credit the increase in the revenue
	Debtors	Asset	AR- Not invoiced	No	Sales-Packing Slip Revenue Offset	X		Golden-1	Debit the increase in the asset

Note-

a. Transaction type above represents the posting profile defined in fig. 1.2

b. 'Cost of Unit delivered' represents the inventory account of the items being shipped based on the posting profile setup. As per my experience, some companies prefer to use a control account during this process instead of a final inventory account. This separates the physical ledger postings from the financial ledger postings for purchase order transactions which is very helpful during inventory reconciliation.

c. The first two lines in the above voucher are always there for all stocked products. The last two lines (in blue) are optional and depend on the flag 'Post to Deferred Revenue Account on Sales Delivery' in the item model group. It will purely be driven by the company policy if the company wants to recognize any revenue upon delivery or wants to book revenue only when it is invoiced.

d. Irrespective of the above flag, all of the entry lines are reversed when the order is invoiced, and a new voucher is generated. So, what happens during packing slip will not make much difference if there is not much time difference between delivery and invoicing.

The accounts that would be configured for this purpose are present under *Cost Management -> Ledger Integration policies setup -> Posting -> Sales Order. See the below picture for accounts that are required to be configured. Fig 1.2*

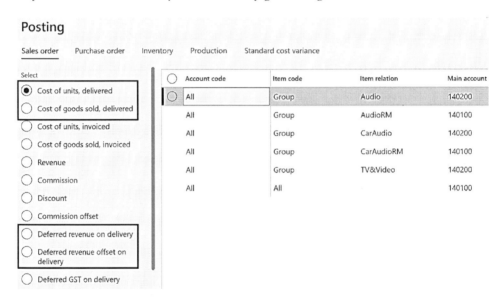

Accounting on shipment/packing of non-stocked products:
Order to Cash has no ledger impact when shipment /packing of services/non-stocked product happens along with stocked products.

4. Invoice Posting

As per Fig 9.1, Invoice posting is another process in Order to Cash cycle where financial transactions are created. Customer Invoice posting is the process that records the company's revenue due to the sale of goods/services and receivables from the customer. This results in an increase in the open customer balance that needs to be collected.

Invoicing through Sales order invoice

Invoicing through the sales order is done for all 'product-related sales, whether the product is an item-type or service-type item.

Accounting on invoicing of stocked products

Below fig. 1.3 shows the accounting that is created when the product receipt for a stocked item is invoiced.

Accounting Trigger in D365	Account Family	Type of Account	Account	Sub ledger impacted	Transaction type in D365	Debit	Credit	Accounting Rule	Logic for Debit/Credit
Invoicing	Debtors	Asset	AR-Invoiced	Yes		X		Golden-1	Debit the increase in the asset
	Sales	Revenue	Sales A/c	No	Sales order revenue		X	Golden-2	Credit the increase in the revenue
	Tax	Tax	Sales Tax/GST Payable A/c	No			X	Golden-1	Credit the increase in the liability
	COGS	Expense	COGS - Finished Goods	No	Cost of goods sold, invoiced	X		Golden-2	Debit the increase in the expense
	Inventory	Asset	Finished Goods Inventory	Yes	Cost of units, invoiced		X	Golden-1	Credit the decrease in the asset
	COGS	Expense	Deferred COGS	No	Cost of Goods Sold, delivered		X	Golden-2	Reversal of packing slip entries
	Inventory	Asset	Finished Goods Inventory	Yes	Cost of Units, delivered	X		Golden-1	
	Sales	Revenue	Accrued Sales (Shipped Not invoiced)	No	Sales-Packing Slip Revenue	X		Golden-2	
	Debtors	Asset	AR- Not invoiced	No	Sales-Packing Slip Revenue Offset		X	Golden-1	

Note-

a. The transaction type above represents the posting profile defined in fig. 1.4

b. Sales Tax/ GST will depend on the local tax law of the country and some countries even don't have any tax on sales.

c. Highlighted lines in fig. 1.3 are the reversal of the packing slip entries as we discussed during the packing slip process as well. So the organizations always make sure that invoicing process takes place immediately after the delivery is done so that all the temporary accounting created during the packing slip are reversed and the final voucher is reflected in the books.

d. In the above example, the 'Finished Goods Inventory' account is used during packing slip and invoicing both. Some organizations create a new WIP account for packing slip purposes to avoid cluttering the 'Finished Goods' ledger account.

Here is a sample voucher entry for the stocked products invoice.

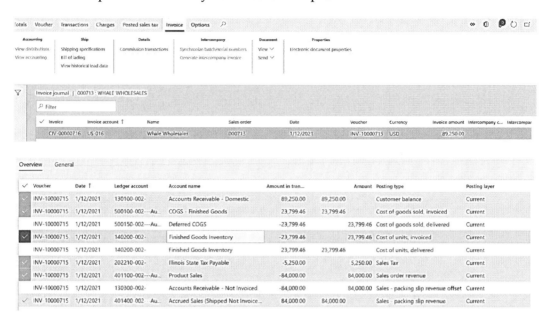

The accounts that would be configured for this purpose are present under *Cost Management -> Ledger Integration policies setup -> Posting -> Sales Order*. See the below picture for accounts that are required to be configured. *Fig. 1.4*

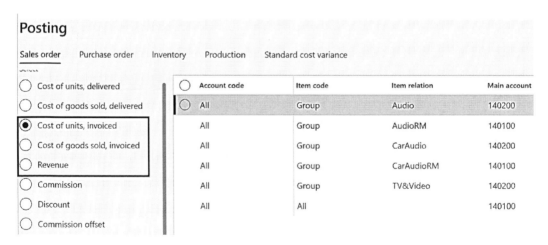

The account setup for Customer balance is present under *Accounts Receivable -> Setup -> Customer posting profiles -> Summary account.*

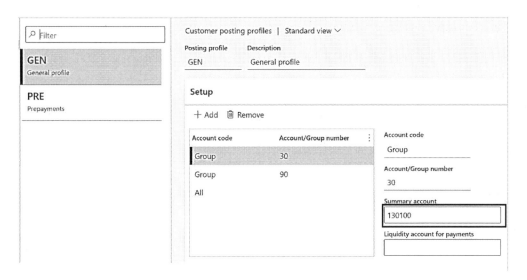

4.1 Accounting on invoicing of non-stocked products

Below are the accounting entries that would occur at the point of invoicing for any non-stocked products or categories are invoiced.

Accounting Trigger in D365	Account Family	Type of Account	Account	Sub ledger impacted	Transaction type in D365	Debit	Credit	Accounting Rule	Logic for Debit/Credit
Invoicing	Debtors	Asset	AR-Invoiced	Yes		X		Golden-1	Debit the increase in the asset
	Sales	Revenue	Sales A/c	No	Sales order revenue		X	Golden-2	Credit the increase in the revenue
	Tax	Tax	Sales Tax/GST Payable A/c	No			X	Golden-1	Credit the increase in the liability

Below is a sample screenshot of accounting entries upon invoicing.

Note-

a. The packing slip is not a required step for non-stocked products, that's why you will not see any reversal entries upon invoicing.

b. Cost of Goods Sold (COGS) is also not calculated for non-stocked products.

The account setup for revenue for non-stocked products is same as stocked products and is present under *Cost Management -> Ledger Integration policies setup -> Posting -> Sales Order.*

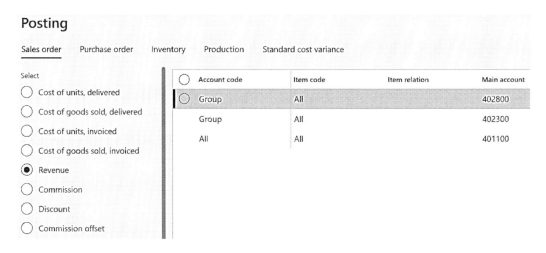

The account setup for Customer balance is present under *Accounts Receivable -> Setup -> Customer posting profiles -> Summary account.*

4.2 Invoicing through Free Text Invoice.

Invoicing through the 'Free Text Invoice' is done for 'non-item related' transactions. It is a very handy feature for those organizations who don't want to create 'non-stocked products for every service they sell. Moreover, some service organizations don't deal in any inventory, so 'free text invoice' is the most important and flexible feature in D365 for them.

Path: *Accounts Receivable* → *Invoices* → *All Free Text Invoices*

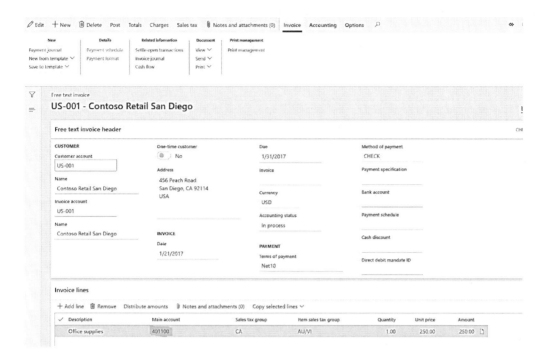

As the name suggests, it is a free text invoice i.e. there is flexibility to select the direct revenue account in this form, unlike sales orders where all the postings are driven by the item posting profile.

5. Process Collection

As per Fig 9.1, customer collection processing is another point in order to cash cycle where financial transactions are created.

Accounting on Customer Collections

Below are the accounting entries that are generated when a customer payment journal is posted:

Accounting Trigger in D365	Account Family	Type of Account	Account	Sub ledger impacted	Transaction type in D365	Debit	Credit	Accounting Rule	Logic for Debit/Credit
Customer Collection	Bank	Asset	Bank A/c	Yes	Bank	X		Golden-1	Debit the increase in the asset
	Debtors	Asset	AR-Invoiced	Yes	Balance		X	Golden-1	Credit the decrease in the asset

Hence, we covered the most important posting types in the order to cash process, but we missed out on some of the options in the 'Sales Order' posting profile which are also commonly used but vary from organization to organization.

Discount Accounting

Most of the organizations do not account for the trade discount but if there is a desire to follow discount accounting, then it is very simple to configure in D365. Essentially, you just need to define the discount account in the 'Discount' type in the sales order posting profile (See below).

Now any sales order which is posted with a discount (percentage or amount) will post the discount amount in the separate discount ledger which is defined above.

Let's take a simple example here. Here is a sales order with a service item and a 20% discount on that.

If there was no discount, the 'product sales' account would have been the net amount. (i.e. 400 in this example). Since we are doing discount accounting, the sales account is credited with the gross amount($500), and the discount is debited with $400. Effectively, the net impact on the profit & loss is still $400 but the accountants can analyze the discount at the ledger level now.

Here is the reasoning behind debit and credit for each line.

Accounting Trigger in D365	Account Family	Type of Account	Account	Sub ledger impacted	Transaction type in D365	Debit	Credit	Accounting Rule	Logic for Debit/Credit
Invoicing with Discount	Debtors	Asset	AR-Invoiced	Yes		429		Golden-1	Debit the increase in the asset
	Sales	Expense	Discount Allowed	No	Discount	100		Golden-2	Debit the increase in the expense
	Tax	Tax	Sales Tax/GST Payable A/c	No			29	Golden-1	Credit the increase in the liability
	Sales	Revenue	Sales A/c	No	Sales order revenue		500	Golden-2	Credit the increase in the revenue

Commission Accounting

Commission programs for the sales team are an important part of any incentive schemes in an organization. D365 has very strong functionality to calculate and post commission for easy commission tracking. There are five set up required for commission calculations e.g.

- Commission Customer group

- Commission Item Group

- Sales Group

- Commission Calculation

- Commission posting

We will not discuss every step here and how to calculate commission but will touch on the posting profile required for commission calculation.

Posting profile for commission can be configured from various paths. There is a small difference between the interface if you access it from the 'sales & marketing' module or access it from the 'Inventory Management' module but it is the same thing in the system.

- Sales & Marketing→ Commission→ Commission posting

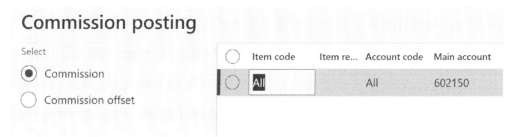

- Inventory Management→ Set up→ Posting → Posting.

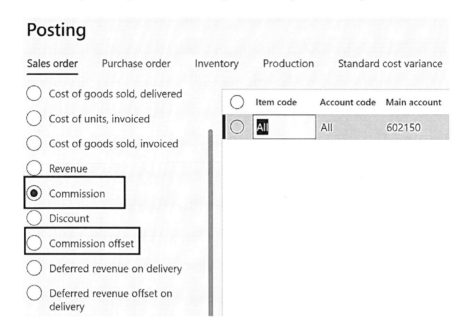

Here is the accounting for commission taking the same example that we discussed during invoice posting. If the commission is configured, it will be just two additional lines in the invoice voucher itself.

Accounting Trigger in D365	Account Family	Type of Account	Account	Sub ledger impacted	Transaction type in D365	Debit	Credit	Accounting Rule	Logic for Debit/Credit
Invoicing	Debtors	Asset	AR-Invoiced	Yes		X		Golden-1	Debit the increase in the asset
	Sales	Revenue	Sales A/c	No	Sales order revenue		X	Golden-2	Credit the increase in the revenue
	Tax	Tax	Sales Tax/GST Payable A/c	No			X	Golden-1	Credit the increase in the liability
	COGS	Expense	COGS - Finished Goods	No	Cost of goods sold, invoiced	X		Golden-2	Debit the increase in the expense
	Inventory	Asset	Finished Goods Inventory	Yes	Cost of units, invoiced		X	Golden-1	Credit the decrease in the asset
	COGS	Expense	Deferred COGS	No	Cost of Goods Sold, delivered		X	Golden-2	Reversal of packing slip entries
	Inventory	Asset	Finished Goods Inventory	Yes	Cost of Units, delivered	X		Golden-1	
	Sales	Revenue	Accrued Sales (Shipped Not invoiced)	No	Sales-Packing Slip Revenue	X		Golden-2	
	Debtors	Asset	AR- Not invoiced	No	Sales-Packing Slip Revenue Offset		X	Golden-1	
	Ledger	Expense	Commission	No	Commission	X		Golden-2	Debit the increase in the expense
	Creditor	Liability	Commission Offset A/c	No	Commission Offset A/c		X	Golden-1	Credit the increase in the liability

10

Inventory Accounting in D365 ERP

Inventory transactions are mainly associated with procurement, sales, manufacturing transactions, and all are controlled through the respective process area in ERP. As we have seen in chapters 7 to 9, inventory accounting is all automated and runs in the background through the very robust posting profile feature of the ERP.

There are some scenarios where the business needs some flexibility and need some easy and quick way to process some transactions or adjustments e.g., adjusting the value of the inventory, adjustment due to month-end stock take, transferring inventory from one bin to another bin, etc. D365 ERP has the following journals which are quite effective in handling such transactions. If you are a financial accountant, you can compare it with the 'General Journal' in the General Ledger module. These journals can be used to bypass certain controls, so not everyone should have access to these.

Following are these journals and the purpose of these journals.

Journal Name	Purpose
Inventory adjustment	This journal is the most popular and used to adjust the inventory quantity and its value.

	The posting accounts are driven by the posting profile.
Movement	Movement journal is similar to adjustment journal, but you can specify the offset account in the movement journal e.g., you can override the contra account. It is widely used to upload the opening balances in a new ERP implementation.
Transfer	This journal is used to transfer items between two stocking locations (warehouses, locations, batches, or product variants). Generally, there is no cost associated with this transfer.
BOM	This inventory journal type is also called mini-production and is very useful in simple or high-volume production scenarios where routes aren't required.
Item arrival	You can use the item arrival journal to register the receipt of items (for example, from purchase orders)
Production input	Production input journals work like the item arrival journals but are used for production orders. (It is not very common)
Counting	Counting journals let you correct the current on-hand inventory that is registered for items or groups of items, and then post the actual physical count, so that you can make the adjustments that are required to reconcile the differences.

10.1 Inventory Adjustment Journal

This journal is the most popular and used to adjust the inventory quantity and its value. The posting accounts are driven by the posting profile. It is widely used to upload the opening balances in a new ERP implementation.

The accounts for this journal and all other inventory-related journals are defined on the following path.

Path: Inventory Management→ Set up→Posting → Posting → Inventory Tab

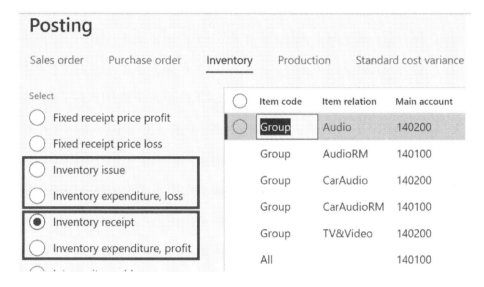

Note:

- 'Inventory issue' and 'Inventory Expenditure, loss' are used together for a scenario where the inventory is adjusted in negative e.g. inventory is written off.

- 'Inventory receipt' and 'Inventory Expenditure, profit' are used together for a scenario where the inventory is adjusted in positive e.g. inventory is written up.

Here is a sample of the movement journal. The item number is selected on one side and there is flexibility to choose the offset account (150100 in this example).

First, you create the journal (aka batch number) and then you can create multiple lines and multiple vouchers in the same batch.

Path: Inventory Management→ Journal Entries→Items→ Inventory Adjustment

There are two levels of posting that take place from any of these journals.

Physical & Financial

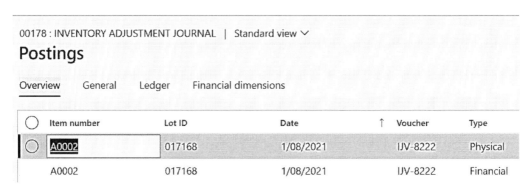

Note- Physical just represents that the posting of such transactions is happening in the inventory books as well.

The financial voucher entry looks like this -

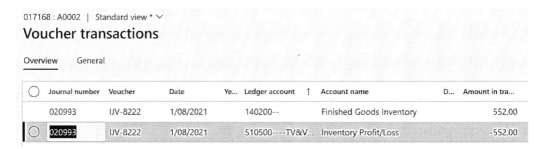

Here is the logic of debit and credit for adjustment journals. In this example, we are increasing the inventory balance, so the offset will be the 'Inventory expenditure, Profit' account.

Accounting Trigger in D365	Account Family	Type of Account	Account	Sub ledger impacted	Transaction type in D365	Debit	Credit	Accounting Rule	Logic for Debit/Credit
Adjustment Journal	Inventory	Asset	Inventory A/c	Yes	Inventory Receipt	X		Golden-1	Debit the increase in the Asset
	Ledger	P&L A/c	Ledger A/c	No	Inventory expenditure, profit		X	Golden-2	Credit the increase in the profit

10.2 Inventory Movement Journal

A movement journal is like an adjustment journal, but you can specify the offset account in the movement journal e.g., you can override the contra account. It is widely used to upload the opening balances in a new ERP implementation.

Here is a sample of the movement journal. The item number is selected on one side and there is flexibility to choose the offset account (150100 in this example).

First, you create the journal (aka batch number) and then you can create multiple lines and multiple vouchers in the same batch.

Path: Inventory Management→ Journal Entries→Items→ Movement

There are two levels of posting that take place from any of these journals.

- Physical

- Financial

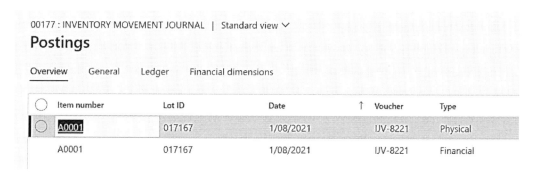

The financial voucher entry looks like this -

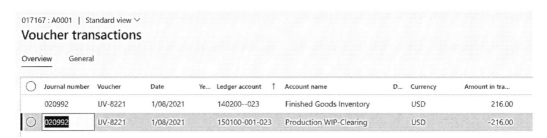

Here is the logic of debit and credit for the movement journal. As you know this journal is a very flexible journal which means you can offset any ledger account (P&L or balance sheet) which can also defy the logic of accounting. In this example, we just picked any clearing account in the offset to balance the transaction which does not make sense and that's the main reason these journals should be given only to people who understand some accounting.

Accounting Trigger in D365	Account Family	Type of Account	Account	Sub ledger impacted	Transaction type in D365	Debit	Credit	Accounting Rule	Logic for Debit/Credit
Movement Journal	Inventory	Asset	Inventory A/c	Yes	Inventory Receipt	X		Golden-1	Debit the increase in the Asset
	Ledger	P&L/ Balance Sheet	Ledger A/c	No	Inventory expenditure, profit		X	Golden-2	Credit the increase in the profit

10.3 Transfer Journal

Movement journal is an altogether different journal, and it is generally used to move inventory from one location to another location.

Path: Inventory Management→ Journal Entries→Items→ Transfer Journal

Here is a sample transaction to transfer 2 qty. of an item from location 11 to location 12.

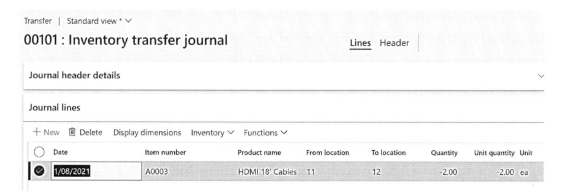

There is no financial entry when you transfer between locations or other stocking locations. There will be just inventory transactions of in and out as you can see in the below screen.

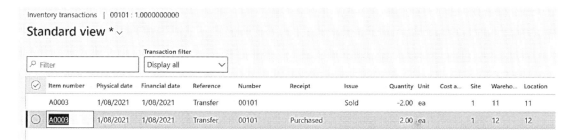

Note: There will be a financial entry on the transfer journal or not, it all depends on the following setup

Path: Product Information Management→ Set up→ Dimension Variant Group→Storage Dimension Group

If the 'financial inventory' option is enabled on the site level (as in this example), it means there will be financial entry only when the transfer journal is used to transfer from site to site. In the above example, we just transferred from a location to another, so it is not tracked financially.

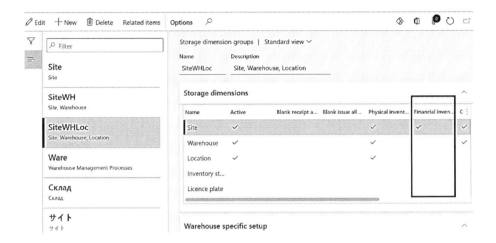

> 💡 *In my experience, generally people track inventory till site level only. Some people also value inventory per warehouse if each warehouse is buying and negotiating their own procurement and the cost to transfer between warehouses is significant.*

10.4 BOM Journal

This inventory journal type is also called mini-production and is very useful in simple or high-volume production scenarios where routes aren't required.

Path: Inventory Management→ Journal Entries→Items→ Bill of Materials

Here is a sample transaction to product 1 qty of F00017 from other 4 different components.

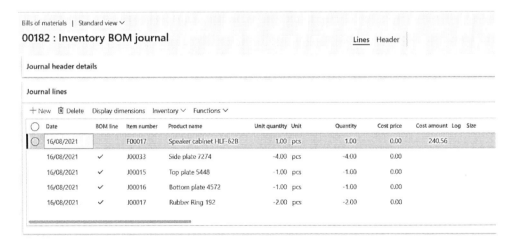

Following is the voucher entry after posting.

Following is the posting profile for defining accounts for finished goods and components consumption

Path: Product Information Management→ Set up→ Posting→ Posting

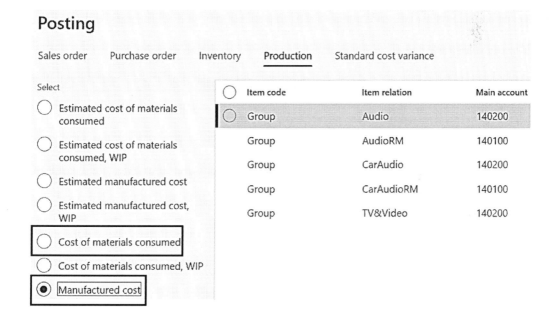

10.5 Item Arrival

You can use the item arrival journal to register the receipt of items (for example, from purchase orders). We will go in the item arrival journals in the Part-II of the book in the series in the chapter on Landed Cost.

10.6 Production Input

Production input journals work like the item arrival journals but are used for production orders. (It is not very common, so we are not going into detail of this)

10.7 Counting

Counting journals let you correct the current on-hand inventory that is registered for items or groups of items, and then post the actual physical count, so that you can make the adjustments that are required to reconcile the differences.

Accounting of the counting journal is the same as 'Inventory Adjustment Journal'

10A

Inventory Valuation & IFRS

We will not talk about how to configure a particular inventory valuation method that is supported in D365, rather we will focus on what are the salient features in each method, what is suitable for a particular industry, and the pros and cons for each of these.

FIFO

FIFO inventory costing technique in simple terms is based on 'First In First Out' i.e. the first receipt is to be settled with the first issue transaction.

> › First in, first-out (FIFO) is an inventory model in which the first acquired receipts are issued first.
> › Financially updated issues from inventory are settled against the first financially updated receipts into inventory, based on the financial date of the inventory transaction.
> › Until month-end transactions are maintained at the average running cost price for the product by storage/tracking dimension

LIFO (Periodic calculation)

> -☼- D365 maintains inventory value at the 'on-hand average' price until the month end is run. 'On-Hand Average' is a running average of the inventory based on the actual purchase prices, so issue transaction is also based on the same average initially.
>
> When the month end is run, the system will value all the issues of inventory using FIFO logic discussed above.

LIFO is an inventory model in which the newest receipts are issued first. Issues from inventory are settled against the last receipts into inventory based on the date of the inventory transaction.

Features:

> › Last in, first-out (LIFO) is an inventory model in which the last (newest) receipts are issued first.
> › Issues from inventory are settled against the last receipts into inventory based on the date of the inventory transaction.
> › If using LIFO Date, if there is no receipt before the issue, the issue is settled against any receipt that occurs after the date of issue.

> -☼- D365 maintains inventory value at the **'on-hand average'** price until the month end is run. 'On-Hand Average' is a running average of the inventory based on the actual purchase prices, so issue transaction is also based on the same average initially.
>
> When the month end is run, the system will value all the issues of inventory

Difference in LIFO and LIFO Date

LIFO works on the basis of whole month as one period i.e. the issues of the month are taken from the last days of the month

Weighted Average

> The weighted average is an inventory model where issues from inventory are valued at the average value of the items that are received into inventory during the inventory closing period, plus any on-hand inventory from the previous period.
> During inventory close, a virtual receipt is created that contains all receipts for the period. The issues are settled against this virtual receipt so that all issues get the same average price.
> Settled using direct or summarized settlement

Month End Application

> At month-end, a summarized weighted average receipt transaction is created. All issues are settled against the weighted average receipt
> Any remaining inventory will be valued at the new calculated weighted average of the receipt transactions
>> • NOTE: If include physical value is selected, Dynamics will use the physical receipts when calculating the running average. Issues are only posted to financially closed transactions during close.

Weighted Average date

> Weighted average dated inventory costing technique in simple terms is a Periodic weighted average principle per day i.e. the issues are valued basis the average of the receipts for every day.
>
> During period close all daily issues are settled against a virtual receipt.
> This allows for all daily issues to have the same average cost.
> The new daily average cost is then used as the inventory cost start for the next day and the calculation is run again
> The same rules apply with direct settlement and use physical cost as the Weighted Average method

Standard Costing

The standard costing technique is straightforward due to its fixed cost approach i.e. all the receipts and issues are valued at the standard cost irrespective of the purchase order prices.

1. With standard costing, the general ledger accounts for inventories and the cost of goods sold contain the *standard costs* of the inputs that should have been used to make the actual good output.

   - Upon PO receipts the inventory is added at the standard cost of the material
   - Upon the 'Report as Finished' (RAF) process, the inventory is added to stock at the standard cost

2. Differences between the actual costs and the standard costs will appear as variances, which can be investigated further.

   - During invoice posting the difference between the standard and actual is posted to the purchase price variance account
   - When ending the production order the difference between the standard and actual is posted to the production variance accounts

3. Analysis of the variance's direct management attention to production inefficiencies or material cost differences.

Moving Average

Moving average technique in simple terms is a point in time cost method (perpetual type) i.e., the issues are always valued basis the point in time cost price on the item.

  › Moving average is a perpetual costing method based upon the average principle where the cost on inventory does not change when the financial purchase cost changes
  › Purchase cost physical vs financial differences are posted to a capitalization account and the remaining amount is expensed.
  › Manual cost revaluations can be completed using today's date only

Comparison of Costing Methods of D365

Method	Pros	Cons
FIFO (First-in-First-out)	• It shows actual inventory costs after the month-end process is run • Sales margin analysis is possible on the run time via marking • Useful in configure/manufacture to order industries	• <u>Requires month-end close process</u> to determine true inventory value • Difficult to analyze for production improvement
LIFO **(Periodic)**	• Rolling true actual inventory costs after month-end • Lower inventory carrying costs using LIFO vs FIFO since the inventory remaining in the books will belong to the earlier purchases which are generally lower cost • Sales margin analysis is possible via marking feature • Useful in configure/manufacture to order industries	• <u>Requires month-end close process</u> to determine true inventory value • Difficult to analyze for production improvement

LIFO **Date**	• Rolling true actual inventory costs after month-end • Lower inventory carrying costs using LIFO vs FIFO since the inventory remaining in the books will belong to the earlier purchases which are generally lower cost • Sales margin analysis is possible via marking feature • Useful in configure/manufacture to order industries	• <u>Requires month-end close process</u> to determine true inventory value • Difficult to analyze for production improvement
Weighted Average **(Periodic)**	• Inventory valued at the average value of receipts at end of every month • Sales margin analysis via marking • Useful in configure/manufacture to order industries	• Requires month-end close process to determine true inventory value • Difficult to analyze for production improvement • Summary settlement transactions can be confusing

Weighted Average **Date**	• Inventory valued at the average value of receipts at end of every month • Sales margin analysis via marking • Useful in configure/manufacture to order industries	• Requires month-end close process to determine true inventory value • Difficult to analyze for production improvement • Summary settlement transactions can be confusing
Standard Costing	• Inventory value is consistent, so there is more predictability of the cost of goods sold. • No month-end close process is required, so there is no adjustment of costing after closing • It helps in the analysis of production/purchasing issues	• Hard to analyze the variances, should not apply to all the industries. • The standard cost roll process is cumbersome to configure in D365
Moving Average	• Rolling true actual inventory cost based upon physical transaction • No month-end close	• Difficult margin analysis • No marking allowed

	• Easy to understand	• No settlements allowed

Which method is best for my business?

It depends on several parameters e.g., location of your business, the inflation rate in the economy or prices are going up or down, and what type of inventory you are dealing with.

Most of the businesses use weighted average or FIFO but many retail businesses also use standard costing as it is always easy for that type of business. Having said that any valuation method is fine if your costing system or ERP system can support that, and it is within the guidelines of IFRS or other standards you follow. There is no one size fits all.

As you must have ready in Chapter-5, there are three methods are allowed as per IAS 2 which are FIFO, Weighted Average and standard costing, and LIFO is specifically prohibited.

I believe this chapter lays a very strong foundation for the consultants in understanding the valuation methods and equip them to recommend a solid approach to the customers.

Visit our website www.satyakejriwal.com for more blogs and videos on Accounting with Microsoft D365.

Section-III
Financial Statements
and Ratio Analysis

11

Financial Statements and Ratio Analysis

"Y̶ou have to understand accounting and you have to understand the nuances of accounting. It's the language of business and it's an imperfect language, but unless you are willing to put in the effort to learn accounting— how to read and interpret financial statements— you really shouldn't select stocks yourself." —**WARREN BUFFETT**

"You have to read a zillion corporate annual reports and their financial statements." —**WARREN BUFFETT**

So, it is very evident if you are value investor, you got to know basic stuff about financial statements. It is like reading a nutrition label of a product, it is easy to learn and understand. If you can cook some food by reading a recipe or apply for a loan by yourself, you can learn the accounting and reading financial statements. It is not rocket science.

There are three main financial statements. They are (1) income statements, (2) balance sheets, (3) cash flow statements. Balance sheets show what a company owns and what it owes at a fixed point in time. Income statements show how much money a company made and spent over a period of time. Cash flow statements show the exchange of money between a company and the outside world also over a period of time.

Let's understand these in detail.

Income Statement

Income statements tell the readers the results of the company's operations for a set period of time. Traditionally, they are reported for each quarter and at the end of the year. Income statements are always shown for the period they cover— such as January 1, 2021, to December 31, 2021. An income statement has the following basic components:

- Revenue

- Cost of Goods Sold (COGS)

- Gross Profit

- Operating Expenses

- EBITDA (Earnings before Income Tax, depreciation & Amortisation)

- Non-Operating Expenses

- Net Profit

Financial Accounting with Microsoft D365 ERP

Following is a sample of an income statement. The format of this statement differs from country to country as per their local requirements but nowadays there are common sections that they report everywhere due to the IFRS requirements which are helping to make the accounting and presentation of financial statements universal globally.

Income Statement for the Period ending 31st Dec, 2021
Contoso Entertainment System USA

	2,021	2,020
Revenue -Stream1	989	887
Revenue -Stream2	636	472
Returns, Refunds, Discounts	(142)	(128)
Total Net Revenue	1,483	1,231
Cost of Goods sold (COGS)	919	790
Gross Profit (GP)	**564**	**441**
Gross Profit %	38%	36%
Selling, General & Admin (SGA)	145	131
Reserarch & Development	80	72
Depreciation	23	21
Total Operating Expenses	**248**	**224**
Earning Before IT, Dep & Amortization (EBITDA)	316	217
EBITDA %	21%	18%
Non-Operating Expenses		
Interest Expense	23	21
Gain/Loss on Sale of Assets & Others	9	8
Income before Tax	284	188
Income Tax Paid	71	64
Net Earning	213	124
Net Profit %	14%	10%

Now, let's divide the statement into three logical sections for analysis:

Income Statement for the Period ending 31st Dec, 2021			
Contoso Entertainment System USA			
		2,021	**2,020**
Revenue -Stream1	Section A	989	887
Revenue -Stream2		636	472
Returns, Refunds, Discounts		(142)	(128)
Total Net Revenue		1,483	1,231
Cost of Goods sold (COGS)		919	790
Gross Profit (GP)		564	441
Gross Profit %		38%	36%
Selling, General & Admin (SGA)	Section B	145	131
Reserarch & Development		80	72
Depreciation		23	21
Total Operating Expenses		248	224
Earning Before IT, Dep & Amortization (EBITDA)		316	217
EBITDA %		21%	18%
Non-Operating Expenses			
Interest Expense	Section C	23	21
Gain/Loss on Sale of Assets & Others		9	8
Income before Tax		284	188
Income Tax Paid		71	64
Net Earning		213	124
Net Profit %		14%	10%

Section A has the sales and gross profit. Sales is also known as the 'Top line' of the company.

		2,021	**2,020**
Revenue -Stream1	Section A	989	887
Revenue -Stream2		636	472
Returns, Refunds, Discounts		(142)	(128)
Total Net Revenue		1,483	1,231
Cost of Goods sold (COGS)		919	790
Gross Profit (GP)		564	441
Gross Profit %		38%	36%

Heading	Explanation	Insight

Revenue	Where the money comes in and it is the main source of income of the company	Commonly called 'Top Line'. The big revenue is good for business but does not give any meaningful indication unless we compare it with expenses
Cost of Goods Sold (COGS)	It is the cost of the main items which are sold by the company and total sales are part of 'Revenue' section above.	This cost is limited to the items sold and does not include any administrative and selling expenses e.g. purchase price of the items and any cost incurred to bring the inventory to the warehouse will be added to the COGS. In ERP, this cost is calculated automatically when you post the sale invoices.
Gross Profit (GP)	GP = Revenue-COGS	
Gross Profit %	GP% = (GP /Revenue) X 100	How much gross profit is good enough for the companies? It varies from industry to industry. **As a general rule, companies with gross profit margins of 40% or more are better and shows a long term competitive advantage but there are exceptions to this rule in some industries.**

Section B is about the operating expenses and EBITDA %

	Section B		
Selling, General & Admin (SGA)		145	131
Reserarch & Development		80	72
Depreciation		23	21
Total Operating Expenses		**248**	**224**
Earning Before IT, Dep & Amortization (EBITDA)		316	217
EBITDA %		21%	18%

Operating Expenses		
Selling, General & Admin (SGA)	It is a major cost of the companies which includes selling expenses, salary,	These expenses can range from **25% to 75% of Gross Profit. If this number is below 25%, those businesses are considered fantastic** but be sure that

	rent, advertising, travel, legal, commission etc.	there are no major expenses under the R&D heading which can offset the competitive advantage of the company
Research & Development	These are expenses like patent, copyright etc. and any expenses incurred on R&D	Companies having big expenses in R&D are sometimes major risk for investors (definitely there are major rewards as well if the research results in some innovation). We should read carefully what are those costs and if those patents etc will generate revenue for a few years and when is that coming to an end and what is the plan after that. **Companies that have to spend heavily on R& D have an inherent flaw in their competitive advantage"**
Depreciation	All machinery and buildings eventually wear out over time; this wearing out is recognized on the income statement as depreciation.	**The amount that something depreciates in a given year is a cost that is allocated against income for that year.** This makes sense: The amount by which the asset depreciated can arguably be said to have been used in the company's business activity of the year that generated the income. Companies that have a competitive advantage tend to have lower depreciation costs as a percentage of gross profit than companies that have to suffer the woes of intense competition.
EBITDA (Earnings Before Income Tax, Depreciation and Amortization)	EBITDA = Gross Profit - SGA-R&D	**Please note that EBITDA does not consider the depreciation cost, so using EBITDA to compare the results of two companies can be misleading sometimes** if those belong to two different industries. Some companies are capital intensive and need large

		capital expenditure and eventually the machines will wear out and the company will have to come up with millions of $ to buy new

Section C is the final section and has the most important result i.e. net profit.

Non-Operating Expenses		
Interest Expense	23	21
Gain/Loss on Sale of Assets & Others	9	8
Income before Tax	284	188
Income Tax Paid	71	64
Net Earning	213	124
Net Profit %	14%	10%

(Section C label overlaps the rows for Interest Expense / Gain/Loss on Sale of Assets & Others)

Non-Operating Expenses		
Interest Expense	This is the interest expense being paid to banks or other lenders for the working capital or long term loans taken by the company.	As a general rule, the interest payment of less than 15% of operating income is considered good. (Just note that this % will be different for the banking industry since earning from the interest difference is their main business)
Gain/Loss on Sale of Assets & Others	Such events would include the sale of fixed assets, such as property, plants, and equipment.	Also included under "Other" would be licensing agreements and the sale of patents if they were categorized as outside the normal course of business.

Income before Tax	**Income before Tax =** Gross Profit- Operating Expenses- Non-Operating Expenses	
Income Tax Paid	It is the income tax being paid by the companies	This is interesting. Sometimes, companies like to tell the world that they are making more money than they are (obvious reasons are the public image of the company for keeping their stock prices rising). You can easily find out if they are showing the correct profit is by the documents they file to Income Tax departments.
Net Earning/Net Profit	Net Profit = Income before Tax - Income Tax Paid	So this is the magic number which is also known as the bottom line.
Net Profit %	Net Profit% = (Net Profit /Revenue) X 100	A simple rule (and there are exceptions) is that if a company is showing a net earnings history of more than 20% on total revenues, there is a real good chance that it is benefiting from some kind of long-term competitive advantage. Likewise, if a company is consistently showing net earnings under 10% on total revenues, you better avoid those companies

Ratio related to Income Statement

Profitability or Sustainability Ratio

Sales Growth	**Sales Growth =** (Current Period Sales –Previous Period Sales)/ Previous Period	Current Year Sale = 1483 Previous Year Sale = 1231	Percentage increase (decrease) in sales between two time periods. Generally, the value investor looks for the

	Sales	Sales Growth = (1483-1231) / 1231 = 20%	companies whose sales are growing year by year at least by 20%
GP Margin	**Gross Profit Margin =** Gross Profit Total Sales/ Total Sales	Gross Profit = 564 Net Sales = 1483 Gross Profit % = 564 / 1483 = 38%	How much profit is earned on your products without considering indirect costs? Is your gross profit margin improving? Small changes in gross margin can significantly affect profitability. Is there enough gross profit to cover your indirect costs? Is there a positive gross margin on all products?
Net Profit Margin	**Net Profit Margin =** Net Profit/ Sales	Net Profit = 213 Net Sales = 1483 Net Profit % = 213 / 1483 = 14%	How much money are you making per every $ of sales? This ratio measures your ability to cover all operating costs including indirect costs
SGA to Sales	**SGA to Sales =** Indirect Costs (sales, general, admin)/ Sales	Indirect Cost = 145 Sales = 1483 SGA to Sales = 145/1483 = 9.77 %	Percentage of indirect costs to sales. Look for a steady or decreasing ratio which means you are controlling overhead
Return on Assets	**Return on Assets =** Net Profit Average/ Total Assets	Net Profit = 213 Total Assets = 1096 Return on Assets % = 213 / 1096 = 19.43%	Measures your ability to turn assets into profit. This is a very useful measure of comparison within an industry.
Return on Equity Ratio	**Return on Equity =** Net Profit / Average Shareholder	Net Profit = 213 Equity= 831.5 Return on Assets % = 213 / 831.5 = 25.61%	Compares capital invested by owners/funders (including grants) and funds provided by lenders. Lenders have priority over

	Equity		equity investors on an enterprise's assets. Lenders want to see that there is some cushion to draw upon in case of financial difficulty.

Balance sheet

Balance sheets are for a particular date. We can create a balance sheet for any day of the year, but it will only be for that specific date. A company's accounting department will generate a balance sheet at the end of each fiscal quarter.

This is a snapshot of the company's financial condition on the particular date that the balance sheet is generated. Now a balance sheet is broken into two parts: T

- The first part is all the assets, and there are many different kinds of assets. They include cash, receivables, inventory, property, plant, and equipment.

- The second part of the balance sheet is liabilities and shareholder equity.

Balance Sheet as on 31st Dec, 2021
Contoso Entertainment System USA

Current Assets	2021	2020
Cash & Cash Equivalents	239.00	211.00
Short Term Investments	3.00	2.00
Receivables	75.00	50.00
Inventory	11.00	10.00
Prepaid Expenses/Other Assets	6.00	5.00
Total Current Assets	334.00	278.00
Long Term Investmets		
Property, Plants & Equipments	338.00	237.00
(-)Accumulated Depreciation	(5.00)	(4.00)
Goodwill	104.00	103.00
Long Term Investmets	325.00	322.00
Other Assets	9.00	8.00
Deferred Long Term Asset	2.00	1.50
Total Assets	1,096.00	936.00
Liabilities		
Current Liabilities		
Accounts payable, other current liabilities	206.00	195.00
Accrues expenses	2.00	1.80
Short Term Debt	3.50	2.30
Total Current Liabilities	211.50	199.10
Long Term Debt	35.00	30.00
Deferred Income Tax	3.00	2.00
Minority Interest	10.00	9.00
Other Liabiliteis	5.00	4.00
Total Liabilities	264.50	244.10
Shareholders Equity		
Issued Capital (Equity) @10/share	150.00	150.00
(-)Treasury Stocks	(30.00)	(30.00)
Retained Earnings	711.50	571.90
Shareholders Equity	831.50	691.90
Total Liabilities & Shareholders Equity	1,096.00	936.00

Balance Sheet as on 31st Dec, 2021
Contoso Entertainment System USA

Current Assets	2021	2020
Cash & Cash Equivalents	239.00	211.00
Short Term Investments Section I	3.00	2.00
Receivables	75.00	50.00
Inventory	11.00	10.00
Prepaid Expenses/Other Assets	6.00	5.00
Total Current Assets	334.00	278.00

Long Term Investmets		
Property, Plants & Equipments	338.00	237.00
(-)Accumulated Depreciation	(5.00)	(4.00)
Goodwill Section II	104.00	103.00
Long Term Investmets	325.00	322.00
Other Assets	9.00	8.00
Deferred Long Term Asset	2.00	1.50
Total Assets	1,096.00	936.00

Liabilities		
Current Liabilities		
Accounts payable, other current liabilities	206.00	195.00
Accrues expenses	2.00	1.80
Short Term Debt	3.50	2.30
Total Current Liabilities	211.50	199.10
Long Term Debt Section III	35.00	30.00
Deferred Income Tax	3.00	2.00
Minority Interest	10.00	9.00
Other Liabiliteis	5.00	4.00
Total Liabilities	264.50	244.10

Shareholders Equity		
Issued Capital (Equity) @10/share	150.00	150.00
(-)Treasury Stocks Section IV	(30.00)	(30.00)
Retained Earnings	711.50	571.90
Shareholders Equity	831.50	691.90
Total Liabilities & Shareholders Equity	1,096.00	936.00

Current Assets	These are called current because these are short term assets i.e. can be converted to cash in a short time (less than a year) Vendor--> Inventory --> Accounts Receivable --> Cash-->Vendor--> Inventory

Cash & Cash Equivalents	It is also known as liquid assets and the companies want to have the right balance of cash and marketable securities.	A huge pile of cash generally means the company's earnings are more than spending. Exceptionally, the reason for cash pile might be (1) issue of new bonds/shared (2) selling a business
Short Term Investments	These are short term deposits or highly liquid bonds	
Receivables	This is the money that is owed to the company by its customers.	If a company is consistently showing a lower percentage of Net Receivables to Gross Sales than its competitors, it usually has some kind of competitive advantage working in its favour that the others don't have.
Inventory	It is a company's product that it buys from its suppliers or manufactures it in the factory and sells it to its customers.	A consultant should always care to see how the inventory is valued in the books. There are different ways of valuation e.g. FIFO, LIFO, Weighted Average etc.
Prepaid Expenses/Other Assets	These are the expenses which are paid in one go but the benefit of that goes for more than a period e.g. insurance paid for 12 months.	
Total Current Assets		

Current Assets			2021	2020
Cash & Cash Equivalents		Section I	239.00	211.00
Short Term Investments			3.00	2.00
Receivables			75.00	50.00
Inventory			11.00	10.00
Prepaid Expenses/Other Assets			6.00	5.00
Total Current Assets			334.00	278.00

Long Term Investmets				
Property, Plants & Equipments		Section II	338.00	237.00
(-)Accumulated Depreciation			(5.00)	(4.00)
Goodwill			104.00	103.00
Long Term Investmets			325.00	322.00
Other Assets			9.00	8.00
Deferred Long Term Asset			2.00	1.50
Total Assets			1,096.00	936.00

Long Term Investments		
Property, Plants & Equipment's	These are the long term assets that are used to produce the goods or ancillary services in the business. These assets are depreciated every year and shown at the net depreciation value. (We already have understood what is depreciation).	As Warren says, producing a consistent product that doesn't have to change equates to consistent profits. The consistent product means there is no need to spend tons of money upgrading the plant and equipment just to stay competitive, which frees up tons of money for other money-making ventures. To get rich, we first have to make money, and it helps if we make lots of money. One of the ways to make lots of money is not having to spend a ton of money keeping up with the Joneses.
Accumulated Depreciation	Depreciation on the assets is shown under the assets as negative number.	

Goodwill / Intangible Assets	When a company buys another business and pays more than its book value, the excess is called Goodwill. This amount is also shown as the net of amortization. Other intangible assets are Patents, Copyrights, Trademarks, franchises etc. This is similar to goodwill in nature and amortized over a period of time.	Whenever we see an increase in goodwill of a company over a number of years, we can assume that it is because the company is out buying other businesses. This can be a good thing if the company is buying businesses that also have growth in the long term.
Long Term Investments	This is the investment that is intended to be longer than one year e.g. stocks, bonds, real estate or can include investment in other businesses or subsidiaries. It is valued at its cost	A company's long-term investments can tell us a lot about the company's thought process. As an investor, we should check the long term investment of the companies, there might be some investment which will unlock value in future.
Other Assets	It includes all other assets which are not included above	
Total Assets	**It is the sum of current assets and long-term assets**	

Liabilities		
Current Liabilities		
Accounts payable, other current liabilities	206.00	195.00
Accrues expenses	2.00	1.80
Short Term Debt	3.50	2.30
Total Current Liabilities	211.50	199.10
Long Term Debt	35.00	30.00
Deferred Income Tax	3.00	2.00
Minority Interest	10.00	9.00
Other Liabiliteis	5.00	4.00
Total Liabilities	264.50	244.10

Section III

Current Liabilities	These are the liabilities which are due in next one year	
Accounts payable, other current liabilities	This is the money that is due to the vendors (Trading) and due within one year.	
Accrues expenses	These are the expenses which are estimated at the end of the year, companies have incurred but invoices are not received yet from the vendors e.g. electricity bill which is received on 5th January, so the accrued expenses on 31st December will be an estimated accrued expenses	It is the result of double-entry system and matching principle of accounting that we have to book expenses to the period where it belongs to.
Short Term Debt	Short term bank loans or credit limits are short term debts.	Short term debts are always expensive than long term debt, so always check what is the ratio of short term and long-term debt. There is no ideal number here but if you see that the short-term debt is rising which shows the management incapability to predict its fund requirement for the

	long term.	
Long Term Debt	These are the liabilities that are due not in this year	Companies that have enough earning power to pay off their long-term debt in under three or four years are considered good companies in long run.
Deferred Income Tax	This is the tax that is accrued but not paid yet.	
Minority Interest	It is the portion of the assets which is not owned by the company.	Minority interest reflects in the balance sheet generally when the company acquires more than 50% (but less than 100%) shares of another company, so it can merge its 100% assets & liabilities in its balance sheet and show in the 'Minority Interest' section what it does not own.
Other Liabilities	The rest of the liabilities fall in this grouping e.g. tax liability, fines etc.	
Total Liabilities		

Shareholders Equity			
Issued Capital (Equity) @10/share	Section IV	150.00	150.00
(-)Treasury Stocks		(30.00)	(30.00)
Retained Earnings		711.50	571.90
Shareholders Equity		831.50	691.90
Total Liabilities & Shareholders Equity		1,096.00	936.00

Shareholders' Equity	Issued Capital (Equity)	This is called equity of the company and shares which are traded in the market. This capital is shown at the originally issued price e.g. $10, $100.	A company can raise new capital by selling bonds or stock (equity) to the public. The money raised by selling bonds has to be paid back at some point in the future. It is borrowed money. But when the company raises money selling preferred or common stock (which is called "equity") to the public, it never has to be paid back. This money is the company's forever, to do with as it pleases.
	Treasury Stocks (-)	These stocks are repurchasing of own stocks from the open market. These are assets (investment) but shown as a deduction from the share capital	When a company buys back its shares, it can do two things with them. It can cancel them or it can retain them with the possibility of reissuing them later on. If it cancels them the shares cease to exist. But if it keeps them, with the possibility of reissuing them later on, they are carried on the balance sheet under shareholders' equity as treasury stock. The presence of treasury shares on the balance sheet are good indicators that the company in question has some advantage working in its favour.
	Retained Earnings	It is that profit of the company which has not been paid out as a dividend but reinvested for the company's growth	Generally, companies use this earning to pay out dividends or buy back their shares from the market, so not having retained earning does not show a bad position always.

Ratio related to Balance sheet

Current Ratio	**Current Ratio =** Current Assets / Current Liabilities	Current Assets = 334 Current Liabilities = 211.5 **Current Ratio = 334/211.5 = 1.58**	It measures your ability to meet short term liability from your short-term assets, which means it is a good indicator of your cash need in near future. A ratio less than 1 may indicate liquidity issues. A very high current ratio may mean there is excess cash that should possibly be invested elsewhere in the business or that there is too much inventory. Most believe that a ratio between 1.2 and 2.0 is sufficient.
Quick Ratio/ Acid Ratio	**Quick Ratio =** Cash +AR + Marketable Securities Current Liabilities	Cash = 239 AR= 75 Securities = 3 Current Liabilities = 11.50 **Quick Ratio = (239+ 75+ 3)/211.50 = 1.5**	A more stringent liquidity test indicates if a firm has enough short-term assets to cover its immediate liabilities. This is often referred to as the "acid test" because it only looks at the company's most liquid assets (excludes inventory) that can be quickly converted to cash). A ratio of 1:1 means that a company can pay its bills without having to sell inventory.
Working Capital	**Working Capital =** Current	Current Assets = 334 Current Liabilities = 211.5	Working Capital is a measure of cash flow and should always be a positive

	Assets – Current Liabilities	**Working Capital =** **334 - 211.5 = 122.5**	number. It measures the amount of capital invested in resources that are subject to quick turnover.

Debt to Equity Ratio	**Debt to Equity Ratio =** Total Debt (short term + Long term) /Equity Capital	Total Debt = 264.50 Equity = 831.50 **Debt to Equity Ratio =** **264.5/831.50 =** **0.32**	Lenders have priority over equity investors on an enterprise's assets. Lenders want to see that there is some cushion to draw upon in case of financial difficulty. The more equity there is, the more likely a lender will be repaid. Most lenders impose limits on the debt/equity ratio, commonly 2:1 for small business loans which means the liability can be maximum double of own equity. In the example here, the debt is 0.32 to 1 which is quite good as the company is mostly relying on its equity.

Most important Numbers (EPS and PE Ratio)
Earnings per Share (EPS)

EPS is earning per share which is net earnings (after tax) divided by the number of outstanding equity shares of the company. This is the earning that is available for the equity stockholders but not necessarily all are distributed among them as dividends since the company retains most of their profit for reinvestment in the company (Which is called retained earnings).

Earning Per Share (EPS)	EPS = Net profit after Tax / Number of Equity Shares	Net Profit = $ 20 m No. of Equity Shares = 10 m EPS = 20 / 10 = $2 per share

PE Ratio (Price to Earning)

It is the most popular ratio among analysts and stock investors to have a good sense of the company's valuation. It is the relationship between a company's stock price and earnings per share (EPS).

P/E Ratio	PE = Share Price / EPS	Share Price (Market value) = $44 EPS= $2 P/E Ratio = $44 / $2 = 22

Earnings are important when valuing a company's stock because investors want to know how profitable a company is and how profitable it will be in the future. Furthermore, if the company doesn't grow and the current level of earnings remains constant, the P/E can be interpreted as the number of years it will take for

Interpreting PE Ratio

Stocks with low PE ratio are considered value stocks i.e. they have high growth potential. In the above example, PE is 22 which means that the price of the share will be recovered in 22 years if we just keep on earning same $ on one share.

It is hard to say if this PE of 22 is very high or low unless we see other companies in the same sector. Generally, there is a grouping done by per sector and their PE ratio is compared. A high PE ratio is also a indication that the stock is overvalued and it is time to consider the exit from that stock.

Value investors like Warren Buffet use this extensively for making their investment decision. They pick stocks with lowest PE ratio and then start analyzing their past profitability trends and estimate future growth based on that.

the company to pay back the amount paid for each share.

Cash Flow

Most of the companies follow the accrual method of accounting as we have already discussed in this book as opposed to a Cash Method. With the accrual method, sales are booked when the goods are shipped, even if the buyer takes years to pay for them.

With a Cash Method, sales are only booked when the cash comes in. Because almost all businesses extend some kind of credit to their buyers, companies have found it more advantageous to use the Accrual Method, which allows them to book the sales on credit as income under accounts receivable on the income statement.

Since the accrual method allows credit sales to be booked as the revenue it has become necessary for companies to keep separate track of the actual cash that flows in and out of the business. To this end, accountants created the cash flow statement. A company can have a lot of cash coming in through the sale of shares or bonds and still not be profitable. (Similarly, a company can be profitable with a lot of sales on credit and not a lot of cash coming in.)

The cash flow statement will tell us only if the company is bringing in more cash than it is spending ("positive cash flow") or if it is spending more cash than it is bringing in ("negative cash flow"). Cash flow statements are like income statements in that they always cover a set period.

The cash flow statement breaks down into three sections:

- Cash flow from operating activities
- Cash flow from Investing activities
- Cash flow from Financing activities

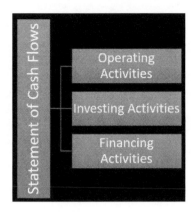

Here is a sample of the cash flow statement and again we have divided it into three sections to understand it a better way :-

Cash Flow - Default

Contoso Entertainment System USA

	December	YTD
Net Income from Operations	**(59,127,229)**	**(23,893,399)**
Add Back Non-Cash Expenses:		
Depreciation and Amortization	85,306	1,008,970
Net Cash Flow from Operations	**(59,041,923)**	**(22,884,429)**
Sources (Uses) of Cash:		
Accounts Receivable	(3,596,380)	(3,890,181)
Liabilities	0	0
Inventory	(7,743,232)	(8,371,168)
Other Assets	0	0
Total Sources (Uses) of Cash	**(11,339,611)**	**(12,261,349)**
Net Cash Flow from Operating Activities	**(70,381,534)**	**(35,145,778)**

	December	YTD
Cash flow from investing activities:		
Additions to Fixed Assets	(19,025,904)	(18,142,440)
	(19,025,904)	(18,142,440)

	December	YTD
Cash flow from financing activities:		
Capital Stock	0	0
Paid-in Capital	0	0
Dividends Paid	0	0
Unrealized Currency Gain/Loss	3,934	7,565
Retained Earnings	0	0
	3,934	7,565
Net Increase (Decrease) in Cash	**(89,403,504)**	**(53,280,653)**
Cash at Beginning of Period	88,810,272	49,409,631
Cash and Equivalents at End of Period	**(593,233)**	**(3,871,022)**

Section A Cash flow from operating activities

Operating activities are revenue-producing activities for the company. Generally, Cash Flow from Operations typically includes the cash flows associated with sales, purchases, and other expenses. There are some expenses where the cash is not involved like depreciation/amortisation, so that is added back to the profit.

Similarly, the change in working capital (change current assets & current liabilities) are also added back which takes out the effect of credit sales from the net income. The reasoning behind this is that all the profit of the company is not cash profit and some part of this comes from credit sales.

Net Income from Operations	**(59,127,229)**	**(23,893,399)**
Add Back Non-Cash Expenses:		
Depreciation and Amortization	85,306	1,008,970
Net Cash Flow from Operations	**(59,041,923)**	**(22,884,429)**
Sources (Uses) of Cash:		
Accounts Receivable	(3,596,380)	(3,890,181)
Liabilities	0	0
Inventory	(7,743,232)	(8,371,168)
Other Assets	0	0
Total Sources (Uses) of Cash	**(11,339,611)**	**(12,261,349)**
Net Cash Flow from Operating Activities	**(70,381,534)**	**(35,145,778)**

Section B Cash flow from investing activities

Investing activities means all the cash flow from acquisition and disposal of non-current assets and other investments e.g. fixed assets purchase and sale or investments purchase or sale.

So, in this section, we will take out all the net additions of fixed assets and investment from the working of section A.

Cash flow from investing activities:		
Additions to Fixed Assets	(19,025,904)	(18,142,440)
	(19,025,904)	(18,142,440)

Section C Cash flow from Financing activities

Cash flow from financing activities:		
Capital Stock	0	0
Paid-in Capital	0	0
Dividends Paid	0	0
Unrealized Currency Gain/Loss	3,934	7,565
Retained Earnings	0	0
	3,934	7,565
Net Increase (Decrease) in Cash	(89,403,504)	(53,280,653)
Cash at Beginning of Period	88,810,272	49,409,631
Cash and Equivalents at End of Period	(593,233)	(3,871,022)

Financing activities means capital borrowing, buying back shares, loan or repayment of loans, dividend payments etc. All the net of these activities are added back or taken out from the above working in section B

Visit our website www.satyakejriwal.com for more blogs and videos on Finance.

12

Financial Statements in D365 ERP

"D365 has some out of the box capabilities for financial statements. There is an inbuilt tool 'financial reports' which was known as 'Management Reporters' before. Microsoft has given some predefined reports there which can be leveraged to build the custom reports as per your requirements quickly.

Financial Reports (Management Reporter)

The tool gets down to the true performance of the business by utilizing General Ledger (GL) data with a drill down into the details if needed and security access is allowed. Sub ledger data should be clean and correct by the time it hits the GL which is why Management Reporter uses this data to create and format boardroom type reports. Management Reporter can also be used for accountants to review and verify that GL data balances and anomalies can be easily researched with the drill-down feature.

- Format your report to contain line numbers to easily identify report rows and statutory formats.
- Choose the financial reporting financial dimensions and attributes that you want in the General ledger. This prevents unnecessary information from showing on your report if it is not critical to financial reporting.
- Print read-only, presentation-ready financial reports to PDF

- View additional details and versions of the financial reports with the report list enhancements. The list of reports shows when the report was last generated and how it was modified. You can also quickly filter your report list by a folder name so you only see the reports relevant to you. View previous versions of the report to accurately track the change in data over time.

- View translated financial data, by any currency in the system, simply by choosing the currency in the report.

- Make it easy for users to see the generated date and time on report headers or footers on a financial report by adding the new auto-text options. Knowing when a report was generated provides crucial information to a user on how new the report data is.

- Analyze data more efficiently with financial dimensions in separate columns when exporting to Microsoft Excel.

- There are a total of 22 predefined reports available out of the box. These will need to be modified to meet your COA and of course, you can create your reports. Clients First can train your users or create the reports for you.

Path : General Ledger → Inquiries and Reports → Financial Reports

Following is the list of reports which are available in D365, we will see these by category in a while.

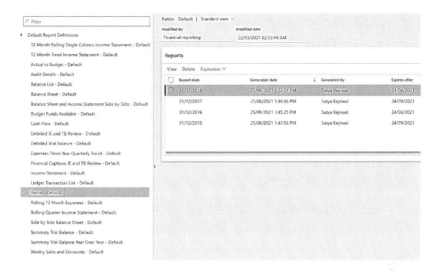

Types of Financial Reports available in D365

These are the categories of reports which are available in D365.

- Trial Balance Reports
- Profit & Loss Reports
- Balance Sheet Reports
- Cash Flow Reports
- Budget Reports
- Ratio Analysis Reports

Trial Balance reports

Following reports are like 'trial balance' reports which means that these reports show only the balances at the ledger level.

1	Summary Trial Balance – Default	View balance information for all accounts that have opening and closing balances, and debit and credit balances together with their net difference.

2	Summary Trial Balance Year Over Year – Default	View balance information for all accounts that have opening and closing balances, and debit and credit balances together with their net difference for the current year and the past year.
3	Detailed Trial Balance - Default	View balance information for all accounts that have debit and credit balances, and the net of these balances, together with the transaction date, voucher, and journal description.
4	Audit Details – Default	View detailed balance information for all accounts. This report shows debit and credit balances in the reporting currency and the local currency, together with additional transaction information, such as the user ID, the user who last modified the data, the date of the last modification, and the journal ID.
5	Balance List – Default	View detailed balance information for all accounts. This report shows opening and closing balances, and debit and credit balances for the current period and year to date, together with additional transaction information, such as the voucher.
6	Detailed JE and TB Review – Default	View opening balance and activity information for all accounts.
7	Ledger Transaction List – Default	View detailed balance information for all accounts. This report shows debit and credit balances, together with additional transaction information, such as the transaction date, journal number, voucher, posting type, and trace number.

Profit & Loss Reports

Following reports are in the nature of 'Profit & Loss' reports and shows the data different ways.

1	Income Statement – Default	View the organization's profitability for the current period and the year to date.
2	2 Month Rolling Single Column Income Statement – Default	View an organization's profitability for the past 12 months in a single column.
3	12 Month Trend Income Statement – Default	View an organization's profitability for each of the last 12 months. These 12 months can span more than one fiscal year.
4	Expenses Three Year Quarterly Trend – Default	Gain insight into expenses for the past 12 quarters over the previous three years.
5	Rolling 12 Month Expenses – Default	Gain insight into expenses for each of the last 12 months. These 12 months can span more than one fiscal year.
6	Rolling Quarter Income Statement – Default	View the organization's profitability on a quarterly basis for the past year and the year to date.
7	Weekly Sales and Discounts - Default	Gain insight into sales and discounts for each week in a month. This report includes a four-week total.

Balance Sheet Reports

Following reports are in the nature of 'Balance Sheet' reports and shows the data in different ways including the comparison from the previous year and period.

1	Balance Sheet – Default	View the organization's financial position for the year.
2	Balance Sheet and Income Statement Side by Side - Default	View the organization's financial position and profitability for the year side by side.
3	Side by Side Balance Sheet – Default	View the organization's financial position for the year. This report shows assets and liability, and shareholder equity side by side.

Cash Flow Reports

There is one standard format available for cash flow report

1	Cash Flow – Default	Gain insight into the cash that is coming in and going out of the organization.

Budget Reports

There are two standard formats available for budgeting reports.

1	Actual vs Budget – Default	View detailed balance information for all accounts for the original budget and compare the revised budget to actuals that have a variance.
2	Budget Funds Available - Default	View a detailed comparison of the revised budget, actual expenditures, budget reservations, and budget funds available for all accounts

Ratio Analysis Reports

Ratios are generally not easy to represent in the reports. D365 has given a basic report for ratio analysis.

1	Ratios – Default	View the solvency, profitability, and efficiency ratios for the organization for the year.

Power BI

But the future belongs to data science and Microsoft has a very solid tool for data visualization which is known as 'Power BI'. Heavy investment is being made in Dataverse by Microsoft which is a data warehouse that can feed data into the reporting tools like Power BI or Tableau

Embedded BI/Dashboard in D365 ERP

Embedded BI is a cut down version of Power BI and helps to create BI dashboards and tiles within ERP. There are more details on the Microsoft website about this.

https://docs.microsoft.com/en-us/power-bi/developer/embedded/

To be cont.....

I am sure you liked this first book in the series. Feel free to give your feedback on our website and if you reckon this book helped you, share this with others as well who can be benefitted by reading this

You can see more such content and videos on our website www.satyakejriwal.com.

And I know there are some of the topics which could not be covered in this book. We are working on the next part which will cover some advanced topics e.g.

- Project Accounting
- Revenue Recognition
- Manufacturing Accounting
- Lease Accounting
- Landed Cost

D365 ERP is expanding every day and Microsoft is making all the effort to include maximum features and functionality in the system e.g., lease accounting and landed cost are the latest edition in the family which we also want to include in our next book.

THANK YOU VERY MUCH for your time.

Acknowledgement

This book is dedicated to God and my father late. Sh. Ramavtar Kejriwal.

I would like to thank my mother (Santosh Kejriwal), my brother (Sandeep Kejriwal), my sisters (Suman, Vandana), sister-in-law (Payal) who have always been a great support for me and helped me achieve all the success in life. My uncles Gopal Bhoot, Kailash Gupta, Subhash Kejriwal, Late. Kishori Lal Kejriwal, Late. Hemant Dadhich for extending the support to the family when we wanted it most.

My wife, life partner and amazing pillar of support Priyanka for tolerating me all these years and never giving up on me. My 10 yrs. old kid (Neil) for keeping a child inside me also alive.

My school- T.I.T. Sr. Sec School in Bhiwani and my first accounting teachers Mr. Singla, Mr. Sukhbir Singh and an inspiring school principal, D.P. Kaushik.

Mr. M.L. Aggarwal whose lecture was the only reason that I used to go to college. His knowledge and fun way of teaching have somehow shaped the way I see accounting now.

Thanks to my friends for the wonderful school and college days memories - Rajesh Agrawal, Manmohan Agrawal, Naveen Mittal, Rakesh Tomar, Anand Singh Verma, Akhilesh Vashishtha, Vikas Agrawal, Sandeep Agrawal, Deepak Vashishtha, Sanjay Gupta, Sunil Athwal, Ganesh Bansal, Rajkumar Dhama

My first employer during CA training (CA. Rishikesh Bhardwaj) and second employer during CA training (CA. N.D. Gupta and CA. Naveen Gupta) for giving me a practical exposure to the business world.

My colleagues and mentors from IL&FS (Mrinal Kapur, Himanshu Mishra), my managers from IL&FS Technologies (Sanjiv Sharma, Sanjiv Bhargava, Namita

Mendiratta) for making me part of the biggest Microsoft ERP implementation of Asia pacific during that time.

My deepest gratitude to our client Mr Rakesh Gupte from IL&FS Mumbai for introducing me to Landmark Education where I learnt "How to live life powerfully and living a life you love". Manu Kochhar, Deepak Pareek, Pawan Kabra, Manish Gupta, D.K. Mittal, Tarvinder Kaur, Ameeta Marfatia for giving me tough business challenges to solve which I tried to solve through Microsoft ERP during implementation. My other clients from my recent years (Lui Sieh and Greg Trigg) showed great confidence in my capability and knowledge.

My utmost gratitude to my ex-boss and CEO of Ignify Inc. (Sandeep Walia) whom I always admire for so many reasons and especially for mentoring people who are working with him.

Special thanks note to the co-author of this book, Anand Singh Verma and other friends who have contributed to this book, Dennis Bacay, Vamsi Pranith, Sandeep Agrawal, Manmohan Agrawal in past. Santosh Pandey for reviewing and giving constructive feedback.

Printed in Great Britain
by Amazon